HOLT
ChemFile
LAB PROGRAM

Consumer Experiments

D

HOLT, RINEHART AND WINSTON
Harcourt Brace & Company

Austin • New York • Orlando • Atlanta • San Francisco • Boston • Dallas • Toronto • London

Contributing Writers: Neo/Sci Corporation

Cover: HRW Photo by Sam Dudgeon

Printed in the United States of America

ISBN 0-03-051882-2

6 179 03 02

LAB MANUAL **D** CONTENTS

CONTENTS continued

Laboratory Safety

HELPING STUDENTS RECOGNIZE THE IMPORTANCE OF SAFETY

One method that can help students appreciate the importance of precautions is to use a "safety contract" that students read and sign, indicating they have read, understand, and will respect the necessary safety procedures, as well as any other written or verbal instructions that will be given in class. You can find a copy of a safety contract on the *Holt ChemFile Teaching Resources CD-ROM.* You can use this form as a model or make your own safety contract for your students with language specific to your lab situation. When making you own contract you could include points such as the following.

- Make sure that students agree to always wear personal protective equipment (goggles and lab aprons). Safety information regarding the use of contact lenses is continually changing. Check your state and local regulations on this subject. Students should agree to read all lab exercises before they come to class. They should agree to follow all directions and safety precautions, and use only materials and equipment that you provide.

- Students should agree to remain alert and cautious at all times in the lab. They should never leave experiments unattended.

- Students should not wear heavy dangling jewelry or bulky clothing.

- Students should bring lab manuals and lab notebooks only into the lab. Backpacks, textbooks for other subjects, and other items elsewhere should be stored elsewhere.

- Students should agree to never eat, drink, or smoke in any science laboratory. Food should NEVER be brought into the laboratory.

- Students should NEVER taste or touch chemicals.

- Students should keep themselves and other objects away from Bunsen-burner flames. Students should be responsible for checking that gas valves and hot plates are turned off before leaving the lab.

- Students should know the proper fire drill procedures and the locations of fire exits.

- Students should always clean all apparatus and work areas.

- Students should wash their hands thoroughly with soap and water before leaving the lab room.

- Students should know the location and operation of all safety equipment in the laboratory.

- Students should report all accidents or close calls to you immediately, no matter how minor.

- Students should NEVER work alone in the laboratory and they should never work unless you are present.

DISPOSAL OF CHEMICALS

Only a relatively small percentage of waste chemicals are classified as hazardous by EPA regulations. The EPA regulations are derived from two acts (as amended) passed by the Congress of the United States: RCRA (Resource Conservation and Recovery Act) and CERCLA (Comprehensive Environmental Response, Compensation, and Liability Act).

In addition, some states have enacted legislation governing the disposal of hazardous wastes that differs to some extent from the federal legislation. The disposal procedures described in this book have been designed to comply with the federal legislation as described in the EPA regulations.

In most cases the disposal procedures indicated in the teacher's edition will *probably* comply with your state's disposal requirements. However, to be sure of this, check with your state's environmental agency. If a particular disposal procedure does not comply with your state requirements, ask that office to assist you in devising a procedure that is in compliance.

The following general practices are recommended in addition to the specific instructions given in the margin notes of this Teacher's Edition.

- Except when otherwise specified in the disposal procedures, neutralize acidic and basic wastes with 1.0 M potassium hydroxide, KOH, or 1.0 M sulfuric acid, H_2SO_4, added slowly while stirring.

- In dealing with a waste-disposal contractor, prepare a complete list of the chemicals you want to dispose of. Classify each chemical on your disposal list as hazardous or nonhazardous waste. Check with your local environmental agency office for the details of such classification.

- Unlabeled bottles are a special problem. They must be identified to the extent that they can be classified as a hazardous or nonhazardous waste. Some landfills will analyze a mystery bottle for a fee if it is shipped to the landfill in a separate package, is labeled as a sample, and includes instructions to analyze the contents sufficiently to allow proper disposal.

ELECTRICAL SAFETY

Although none of the labs in this manual require electrical equipment, several include options for the use of microcomputer-based laboratory equipment, pH meters, or other equipment. The following safety precautions to avoid electric shocks must be observed any time electrical equipment is present in the lab.

- Each electrical socket in the laboratory must be a three-hole socket and must be protected with a GFI (ground-fault interrupter) circuit.

- Check the polarity of all circuits before use with a polarity tester from an electronics supply store. Repair any incorrectly wired sockets.

- Use only electrical equipment equipped with a three-wire cord and three-prong plug.

- Be sure all electrical equipment is turned off before it is plugged into a socket. Turn off electrical equipment before it is unplugged.

- Wiring hookups should be made or altered only when apparatus is disconnected from the power source and the power switch is turned off.

- Do not let electrical cords dangle from work stations; dangling cords are a shock hazard and students can trip over them.

- Do not use electrical equipment with frayed or twisted cords.

- The area under and around electrical equipment should be dry; cords should not lie in puddles of spilled liquid.

- Hands should be dry when using electrical equipment.

- Do not use electrical equipment powered by 110–115 V alternating current for conductivity demonstrations or for any other use in which bare wires are exposed, even if the current is connected to a lower voltage AC or DC connection.

Use dry cells or Ni-Cad rechargeable batteries as direct current sources. Do not use automobile storage batteries or AC-to-DC converters; these two sources of DC current can present serious shock hazards.

Prepared by Jay A. Young, Consultant, Chemical Health and Safety, Silver Spring, Maryland

Master Materials List

All experiments were bench-tested by the technical staff at Neo/Sci Corporation to ensure a successful lab experience. Neo/Sci has also prepared the following materials list to aid in ordering supplies for this manual. Items are listed alphabetically for each category. Each item is followed by a Neo/Sci product number. Amounts specified are for 15 lab groups.

Supplies can be ordered and technical questions can be answered about any experiment in this manual by contacting Neo/Sci at the following:

Neo/Sci Corporation
150 Lucius Gordon Drive #114
P.O. Box 22729
Rochester, NY 14692-2729
Toll-free phone: 1-800-526-6689
Toll-free fax: 1-800-657-7523

Chemicals

Chemical	Product number	Amount	Experiment
Acetic acid, glacial	H9-95-0130	5.8 mL	D17
Alizarin	H9-95-0140	2 g	D6
Allura red	H9-95-0150	2 g	D6
Ammonia	H9-95-0160	100 mL	D17
Benedict's solution, Qual.	H9-95-0220	825 mL	D1, D5
Bismarck Brown Y	H9-95-0225	2 g	D6
Bleach	H9-95-0231	1 gal	D19
Borax	H9-95-0233	8 g	D9
Boric acid	H9-95-0232	6.2 g	D17
Bromothymol blue	H9-95-0235	15 mL	D15
Calcium carbonate	H9-95-0305	40 g	D13
Calcium hydroxide	H9-95-0315	70 g	D3
Chymosin	H9-95-0325	2 mL	D21
Ethyl alcohol, 95%	H9-95-0530	695 mL	D7, D16
Ferric chloride, $FeCl_3$	H9-95-0605	1 g	D5
Guar gum	H9-95-0750	17 g	D9
Hydrochloric acid, conc.	H9-95-0814	221.5 mL	D3, D5, D15, D17, D20
India ink	H9-95-0920	2 mL	D18
Iodine crystals	H9-95-0930	7 g	D7
Iodine, tincture	H9-95-0935	45 mL	D7
Isopropyl alcohol, denatured	H9-95-0940	10 mL	D13
Lemon juice	H9-95-1230	120 mL	D14, D21

Chemical	Product number	Amount	Experiment
Lugol's iodine	H9-95-1250	45 mL	D5
Magnesium sulfate	H9-95-1305	1650 g	D12
Methyl orange	H9-95-1325	2 g	D6
Milk of magnesia	H9-95-1335	15 mL	D17
Petroleum jelly	H9-95-1605	13 oz	D19
Phenolphthalein	H9-95-1612	15 mL	D16
Polyvinyl alcohol	H9-95-1625	50 g	D9
Potassium iodide	H9-95-1631	5 g	D7
Potassium sulfate	H9-95-1634	237 g	D6
Rennet tablets	H9-95-1840	1	D21
Rust inhibitor	H9-95-1860	1	D19
Sodium bicarbonate	H9-95-1975	8.4 g	D17
Sodium carbonate	H9-95-1977	286 g	D5, D6, D11
Sodium chloride	H9-95-1920	200 g	D8, D17, D19
Sodium hydroxide	H9-95-1983	108 g	D6, D15, D16, D17
Sodium polyacrylate	H9-95-1985	18 g	D2, D8
Sucrose	H9-95-1991	390 g	D11
Sudan IV	H9-95-1992	0.5 g	D13
Sunset yellow	H9-95-1994	2 g	D6
Vinegar, white	H9-95-2210	665 mL	D11, D19
Yeast, baker's	H9-95-2530	225 g	D11

Equipment

Equipment	Product number	Amount	Experiment
Balance	H9-55-0050	15	D1, D2, D5, D8, D10, D11, D13, D19, D20, D21
Beaker tongs	H9-60-0150	15	D7
Beaker, 1 L	H9-60-0135	45	D6, D9
Beaker, 100 mL	H9-60-0110	45	D19
Beaker, 150 mL	H9-60-0115	75	D12, D15, D18
Beaker, 25 mL	H9-60-0100	90	D7
Beaker, 250 mL	H9-60-0120	45	D3, D10, D11, D15, D17, D21
Beaker, 400 mL	H9-60-0125	15	D12
Beaker, 50 mL	H9-60-0105	30	D21
Beaker, 500 mL	H9-60-0130	45	D1, D5, D7, D13 D19, D20

Equipment	Product number	Amount	Experiment
Buret clamp, double	H9-60-0155	15	D15, D16
Buret, 50 mL	H9-60-0160	15	D15, D16
Cork stopper	H9-60-3010	15	D10
Dispensing bottle, 30 mL	H9-99-0005	6	D17
Erlenmeyer flask 125 mL	H9-60-1005	75	D13
Erlenmeyer flask, 250 mL	H9-60-1010	135	D11, D16
Forceps	H9-60-0165	15	D17, D19
Funnel	H9-60-0170	15	D9, D16, D21
Glass tubing, 4 mm	H9-60-2000	15	D3
Graduated cylinder, 10 mL	H9-60-1050	90	D5, D8, D15, D21
Graduated cylinder, 100 mL	H9-60-1065	15	D1, D2, D3, D8, D10, D13, D16, D18
Graduated cylinder, 25 mL	H9-60-1055	75	D5, D7
Graduated cylinder, 50 mL	H9-60-1060	60	D9, D19, D21
Hot plate	H9-55-0070	15	D1, D5, D6, D7, D9, D13, D14, D17, D21
Litmus paper, neutral	H9-99-8015	300 strips	D17
Magnifying glass	H9-99-8040	15	D3, D6, D20
Medicine dropper	H9-99-0203	30	D2, D5, D13, D17, D18, D21
Metal file	H9-60-0180	15	D10
Micropipet	H9-60-0010	15	D8
Microspatula	H9-60-0015	15	D2, D5, D8, D15
Mortar and pestle	H9-60-2040	15	D5, D15, D20
Pizza cutter	Local	15	D4
pH meter	H9-55-2025	15	D15
pH paper, 1-14 range	H9-99-8001	7 rolls	D3, D5, D11, D14, D17, D18, D21
Razor blade	H9-60-0175	15	D10
Ring stand	H9-60-0186	15	D1, D7, D9, D15, D16
Ring, iron	H9-60-0185	15	D1, D9
Rubber stopper, #2 1-hole	H9-60-3021	15	D3
Rubber stopper, #3	H9-60-3030	15	D18
Rubber stopper, #5	H9-60-3050	15	D9, D13
Rubber tubing	H9-60-3110	15	D3
Ruler, clear 6"	H9-98-0020	15	D9, D10, D13, D17, D19, D22

Spatula	H9-60-0020	15	D7, D9, D14
Spot plate, 6-well	H9-99-0230	15	D17
Stir rod, glass	H9-99-0235	15	D13, D15, D20
Stopwatch	H9-65-2045	15	D18, D21
Test tube clamp	H9-60-0153	15	D1, D3, D5
Test tube rack	H9-60-0190	15	D1, D3, D5, D7
Test tube, 16 × 125 mm	H9-99-0253	30	D19
Test tube, 20 × 150 mm	H9-99-0254	150	D1, D3, D5, D7, D20
Thermometer clamp	H9-60-0154	15	D7
Thermometer, alcohol −10 to 120°C	H9-60-2050	150	D7, D9, D10, D11, D13, D21
Thermometer, candy	Local	15	D14
Utility tongs	H9-60-0151	15	D6
Watch glass	H9-60-2045	15	D21
Weighing paper	H9-99-8100	30	D2, D8

Miscellaneous

Miscellaneous	Product number	Amount	Experiment
Aluminum foil pan, oblong	Local	15	D21
Aluminum foil, heavy	H9-60-3200	1	D10
Aluminum foil, light	H9-60-3201	1	D12
Aluminum pie pan	Local	30	D9, D10
Antacid, liquid, 5 brands	Local	75 mL each	D15
Antacid, tablets, 5 brands	Local	75 each	D15
Apple	Local	1	D17
Aspirin, tablets, 5 brands	Local	150 each	D5, D16
Balloons, round	H9-98-6150	135	D11
Bar magnet	H9-99-6100	15	D20
Bar soap, 5 brands	Local	1 g each	D13
Butter	Local	15 T	D7
Can opener	Local	15	D10
Carbonated beverage, 5 brands	Local	1 each	D1
Cardstock 6 × 2.5 in.	Local	15	D9
Cereal with iron, 3 brands	Local	15 each	D20
Cheesecloth	H9-98-6072	8 yards	D14, D21
Chocolate, milk 1 × 0.25 in. piece	Local	15	D7
Clothesline	Local	1	D6, D17

Miscellaneous	Product number	Amount	Experiment
Clothespins	Local	90	D6, D17
Coconut oil	Local	150 mL	D7
Cod liver oil	Local	150 mL	D7
Colander	Local	15	D14
Comb, nylon	Local	15	D22
Copper BBs	H9-98-6175	75	D18
Corn oil	Local	225 mL	D7
Cornstarch	H9-95-0330	50 g	D9
Cotton fabric	H9-98-6074	90	D6
Dishpan, plastic	Local	15	D6
Dowel, wood 1/4" diameter	Local	15	D9
Duct Tape	Local	1 roll	D9
Egg, fresh	Local	1-2	D17
Fabric squares 2 × 2 in.	Local	15	D13
Filter paper, 9 cm circle	H9-99-8018	90	D17
Food coloring, 4 colors	H9-95-0620	2 mL each	D9
Fruit jelly	Local	15 mL	D17
Galvanized nails	Local	45	D19
Ginger ale	Local	15 mL	D17
Grapefruit	Local	1	D17
Hot mitt	Local	15	D4, D10, D17
Ice	Local		D9, D11
Incubator	H9-55-0080	15	D11
Iron nails	Local	75	D19
Lard	Local	5 T	D13
Laundry marker	Local	15	D6
Lens tissue	H9-99-8019	1	D3
Liquid detergent, 5 brands	Local	300 mL ea.	D13
Liquid soap, 5 brands	Local	300 mL ea.	D13
Margarine, soft	Local	15 T	D7
Margarine, stick	Local	15 T	D7
Matches	Local	15 books	D10
Medicine cup, 1 oz	H9-98-3150	225	D5
Microscope slide, plastic	H9-60-2055	150	D3
Milk, whole	Local	34 L	D14, D17, D21
Mineral water	Local	15 mL	D17
Molasses	Local	15 mL	D17
Nail polish, clear	Local	1	D19

ChemFile Lab D

Miscellaneous	Product number	Amount	Experiment
Orange	Local	1-2	D17
Oven	Local	8	D4
Paper clips	H9-98-0040	123	D10, D12, D19
Paper cup, 2 oz	H9-98-3102	75	D3
Paper towels	Local	1 roll	D19, D20
Peanut oil	Local	150 mL	D7
Peanut, halves	Local	45	D10
Pencils, colored	H9-98-0012	15	D17
Pennies	Local	45	D19
Petri dish	H9-99-0220	105	D8
Pizza mix	Local	15	D4
Pizza pan	Local	15	D4
Plastic cup, 4 oz clear	H9-98-3104	195	D2, D9
Plastic wrap	Local	1 piece	D9
Popsicle sticks	H9-98-6160	30	D21
Powdered detergent, 5 brands	Local	60 g each	D6, D13
Quarters	Local	45	D19
Red cabbage	Local	1-2	D17
Rubberbands	H9-98-0016	135	D11
Saucepan, 2 qt	Local	15	D14
Sauerkraut juice	Local	15 mL	D17
Scissors, fine point	H9-99-8140	15	D17
Shampoo, clear, 5 brands	Local	30 mL each	D18
Shredded paper	Local	15 g	D19
Steel wool	H9-99-6240	15 g	D19
String, cotton	Local	2 m	D12, D17
Sunflower oil	Local	150 mL	D7
Sweet potato	Local	1-2	D17
Tablespoon	Local	15	D7, D14
Tape measure	Local	15	D11
Tape, transparent	Local	1 roll	D9, D20
Tin can 16 oz	Local	15	D10
Tin snips	Local	15	D10
Tomato	Local	1-2	D17
Toothpaste, 5 brands	Local	1 each	D3
Toothpicks	H9-98-6165	75	D3
Vegetable shortening	Local	15 T	D7
Walnut halves	Local	45	D10

Miscellaneous	Product number	Amount	Experiment
Wash bottle, 100 mL	H9-60-1080	15	D16
Water, bottled 5 brands	Local	1 each	D2
Wax paper	Local	1 roll	D13
Wax pencil	H9-98-0010	15	D1, D2, D3, D5, D7, D8, D9, D11, D13, D15, D17, D18, D19, D20, D21
White glue	Local	1.02 L	D6, D9

Safety Equipment

Safety Equipment	Product number	Amount	Experiment
Apron, lab	H9-60-4010	30	D1, D2, D3, D4, D5, D6, D7, D8, D9, D10, D11, D12, D13, D14, D15, D16, D17, D18, D19, D20, D21, D22
Face shield	H9-60-4000	1	
Gloves, impermeable	H9-60-4005	1	
Goggles, safety	H9-60-4015	30	D1, D2, D3, D4, D5, D6, D7, D8, D9, D10, D11, D12, D13, D14, D15, D16, D17, D18, D19, D20, D21, D22

Neo/Sci Lab Kits

(each kit contains enough materials for 15 lab groups)

Experiment	Product number	Amount
D3 A Close Look at Toothpaste	H9-20-1783	1
D5 A Close Look at Aspirin	H9-20-1773	1
D8 Polymers as Straws	H9-20-1733	1
D9 The Slime Challenge	H9-20-1753	1
D11 Factors Affecting CO_2 Production in Yeast	H9-20-1343	1
D12 Solutions: Rock Formation	H9-20-1533	1
D13 A Close Look at Soaps and Detergents	H9-20-1803	1
D15 How Effective Are Antacids?	H9-20-1793	1
D18 Shampoo Chemistry	H9-20-1763	1
D19 Rust Race	H9-20-1713	1
D21 Curdling the Bio-Tech Way	H9-20-1453	1

Chemical Inventory

There are many computer programs and data bases that you can use to track your chemical inventory. The following is an example of the information you could include within your inventory tracking system.

School Name: _____

Street Address: _____

City/County/Zip Code: _____

Chemical name and concentration or form	Estimated amount on hand	Amount and date purchased	Comments
Acetic acid, 5% solution	250 mL	2 L 9/97	Can substitute white vinegar
Acetic acid, glacial	900 mL	2 L 10/96	
$AgNO_3$, solid, technical grade	300 g	500 g 4/95	
$AgNO_3$, 1 M solution	200 mL	500 mL 9/97	Store in amber bottles
$Ba(NO_3)_2$, solid, technical grade	40 g	100 g 4/95	
$BaCl_2$, anhydrous form	130 g	250 g 4/95	
$BaCl_2 \cdot 5H_2O$, solid	15 g	100 g 4/95	Can substitute the anhydrous form

Incompatible Chemicals

The following listing should be considered when organizing and storing chemicals. Note that some chemicals on this list should no longer be in your lab due to their potential health risks. Consult your state or local guidelines for more specific information on chemical hazards.

Chemical	Should not come in contact with
Acetic acid	Chromic acid, nitric acid, perchloric acid, ethylene glycol, hydroxyl compounds, peroxides, and permanganates
Acetone	Concentrated sulfuric acid and nitric acid mixtures
Acetylene	Silver, mercury and their compounds; bromine, chlorine, fluorine, and copper tubing,
Alkali metals, powdered aluminum and magnesium	Water, carbon dioxide, carbon tetrachloride, and the halogens

Chemical	Should not come in contact with
Ammonia (anhydrous)	Mercury, hydrogen fluoride, and calcium hypochlorite
Ammonium nitrate (strong oxidizer)	Strong acids, metal powders, chlorates, sulfur, flammable liquids, and finely-divided organic materials
Aniline	Nitric acid and hydrogen peroxide
Bromine	Ammonia, acetylene, butane, hydrogen, sodium carbide, turpentine, and finely-divided metals
Carbon (activated)	Calcium hypochlorite, all oxidizing agents
Chlorates	Ammonium salts, strong acids, powdered metals, sulfur, and finely-divided organic materials
Chromic acid	Glacial acetic acid, camphor, glycerin, naphthalene, turpentine, low-molar mass alcohols, and flammable liquids
Chlorine	Same as bromine
Copper	Acetylene and hydrogen peroxide
Flammable liquids	Ammonium nitrate, chromic acid, hydrogen peroxide, sodium peroxide, nitric acid, and the halogens
Hydrocarbons (butane, propane, gasoline, turpentine)	Fluorine, chlorine, bromine, chromic acid, and sodium peroxide
Hydrofluoric acid	Ammonia
Hydrogen peroxide	Most metals and their salts, flammable liquids, and other combustible materials
Hydrogen sulfide	Nitric acid and certain other oxidizing gases
Iodine	Acetylene and ammonia
Nitric acid	Glacial acetic acid, chromic and hydrocyanic acids, hydrogen sulfide, flammable liquids, and flammable gases that are easily nitrated
Oxygen	Oils, grease, hydrogen, flammable substances
Perchloric acid	Acetic anhydride, bismuth and its alloys, alcohols, paper, wood, and other organic materials
Phosphorus pentoxide	Water
Potassium permanganate	Glycerin, ethylene glycol, and sulfuric acid
Silver	Acetylene, ammonium compounds, oxalic acid, and tartaric acid
Sodium peroxide	Glacial acetic acid, acetic anhydride, methanol, carbon disulfide, glycerin, benzaldehyde, and water
Sulfuric acid	Chlorates, perchlorates, permanganates, and water

ChemFile LAB D

Consumer Experiments

HOLT, RINEHART AND WINSTON
Harcourt Brace & Company

Austin • New York • Orlando • Atlanta • San Francisco • Boston • Dallas • Toronto • London

Contributing Writers: Neo/Sci Corporation

Cover: HRW Photo by Sam Dudgeon

Printed in the United States of America

ISBN 0-03-054374-6

6 179 03 02

Introduction to the Lab Program

STRUCTURE OF THE EXPERIMENTS

INTRODUCTION
The opening paragraphs set the theme for the experiment and summarize its major concepts.

OBJECTIVES
Objectives highlight the key concepts to be learned in the experiment and emphasize the science process skills and techniques of scientific inquiry.

MATERIALS
These lists enable you to organize all apparatus and materials needed to perform the experiment. Knowing the concentrations of solutions is vital. You often need this information to perform calculations and to answer the questions at the end of the experiment.

SAFETY
Safety cautions are placed at the beginning of the experiment to alert you to procedures that may require special care. Before you begin, you should review with the safety issues that apply to the experiment.

PROCEDURE
By following the procedures of an experiment, you are performing concrete laboratory operations that duplicate the fact-gathering techniques used by professional chemists. You are learning skills in the laboratory. The procedures tell you how and where to record observations and data.

DATA AND CALCULATIONS TABLES
The data that you will collect during each experiment should be recorded in the labeled Data Tables provided. The entries you make in a Calculations Table emphasize the mathematical, physical, and chemical relationships that exist among the accumulated data. Both tables should help them to think logically and to formulate their conclusions about what occurred during the experiment.

CALCULATIONS
Space is provided for all computations based on the data you have gathered.

QUESTIONS
Based on the data and calculations, you should be able to develop a plausible explanation for the phenomena observed during the experiment. Specific questions are asked that require you to draw on the concepts you have learned.

GENERAL CONCLUSIONS
This section asks broader questions that bring together the results and conclusions of the experiment and relate them to other situations.

Safety in the Chemistry Laboratory

Chemicals are not toys

Any chemical can be dangerous if it is misused. Always follow the instructions for the experiment. Pay close attention to the safety notes. Do not do anything differently unless told to do so by your teacher.

Chemicals, even water, can cause harm. The trick is to know how to use chemicals correctly so that they will not cause harm. If you follow the rules stated in the following pages, pay attention to your teacher's directions, and follow the cautions on chemical labels and the experiments, then you will be using chemicals correctly.

These safety rules always apply in the lab

1. **Always wear a lab apron and safety goggles.**
   Even if you aren't working on an experiment, laboratories contain chemicals that can damage your clothing, so wear your apron and keep the strings of the apron tied. Because chemicals can cause eye damage, even blindness, you must wear safety goggles. If your safety goggles are uncomfortable or get clouded up, ask your teacher for help. Try lengthening the strap a bit, washing the goggles with soap and warm water, or using an antifog spray.

2. **No contact lenses are allowed in the lab.**
   Even while wearing safety goggles, chemicals could get between contact lenses and your eyes and cause irreparable eye damage. If your doctor requires that you wear contact lenses instead of glasses, then you should wear eye-cup safety goggles in the lab. Ask your doctor or your teacher how to use this very important and special eye protection.

3. **Never work alone in the laboratory.**
   You should always do lab work only under the supervision of your teacher.

4. **Wear the right clothing for lab work.**
   Necklaces, neckties, dangling jewelry, long hair, and loose clothing can cause you to knock things over or catch items on fire. Tuck in neckties or take them off. Do not wear a necklace or other dangling jewelry, including hanging earrings. It isn't necessary, but it might be a good idea to remove your wristwatch so that it is not damaged by a chemical splash.

   Pull back long hair, and tie it in place. Nylon and polyester fabrics burn and melt more readily than cotton, so wear cotton clothing if you can. It's best to wear fitted garments, but if your clothing is loose or baggy, tuck it in or tie it back so that it does not get in the way or catch on fire.

   Wear shoes that will protect your feet from chemical spills—no open-toed shoes or sandals and no shoes with woven leather straps. Shoes made of solid leather or a polymer are much better than shoes made of cloth. Also, wear pants, not shorts or skirts.

5. **Only books and notebooks needed for the experiment should be in the lab.**
   Do not bring other textbooks, purses, bookbags, backpacks, or other items into the lab; keep these things in your desk or locker.

6. **Read the entire experiment before entering the lab.**
   Memorize the safety precautions. Be familiar with the instructions for the experiment. Only materials and equipment authorized by your teacher should

be used. When you do the lab work, follow the instructions and the safety precautions described in the directions for the experiment.

7. **Read chemical labels.**
Follow the instructions and safety precautions stated on the labels. Know the location of Materials Safety Data Sheets for chemicals.

8. **Walk carefully in the lab.**
Sometimes you will carry chemicals from the supply station to your lab station. Avoid bumping other students and spilling the chemicals. Stay at your lab station at other times.

9. **Food, beverages, chewing gum, cosmetics, and smoking are NEVER allowed in the lab.**
You already know this.

10. **Never taste chemicals or touch them with your bare hands.**
Also, keep your hands away from your face and mouth while working, even if you are wearing gloves.

11. **Use a sparker to light a Bunsen burner.**
Do not use matches. Be sure that all gas valves are turned off and that all hot plates are turned off and unplugged when you leave the lab.

12. **Be careful with hot plates, Bunsen burners, and other heat sources.**
Keep your body and clothing away from flames. Do not touch a hot plate after it has just been turned off. It is probably hotter than you think. The same is true of glassware, crucibles, and other things after you remove them from a hot plate, drying oven, or the flame of a Bunsen burner.

13. **Do not use electrical equipment with frayed or twisted cords or wires.**

14. **Be sure your hands are dry before using electrical equipment**
Before plugging an electrical cord into a socket, be sure the electrical equipment is turned off. When you are finished with it, turn it off. Before you leave the lab, unplug it, but be sure to turn it off first.

15. **Do not let electrical cords dangle from work stations; dangling cords can cause tripping or electrical shocks.**
The area under and around electrical equipment should be dry; cords should not lie in puddles of spilled liquid.

16. **Know fire drill procedures and the locations of exits.**

17. **Know the location and operation of safety showers and eyewash stations.**

18. **If your clothes catch on fire, walk to the safety shower, stand under it, and turn it on.**

19. **If you get a chemical in your eyes, walk immediately to the eyewash station, turn it on, and lower your head so that your eyes are in the running water.**
Hold your eyelids open with your thumbs and fingers, and roll your eyeballs around. You have to flush your eyes continuously for at least 15 mm. Call your teacher while you are doing this.

20. **If you have a spill on the floor or lab bench, call your teacher rather than trying to clean it up by yourself.**
Your teacher will tell you if it is OK for you to do the cleanup; if it is not, your teacher will know how the spill should be cleaned up safely.

21. **If you spill a chemical on your skin, wash it off under the sink faucet, and call your teacher.**

    If you spill a solid chemical on your clothing, brush it off carefully so that you do not scatter it, and call your teacher. If you get a liquid on your clothing, wash it off right away if you can get it under the sink faucet, and call your teacher. If the spill is on clothing that will not fit under the sink faucet, use the safety shower. Remove the affected clothing while under the shower, and call your teacher. (It may be temporarily embarrassing to remove your clothing in front of your class, but failing to flush that chemical off your skin could cause permanent damage.)

22. **The best way to prevent an accident is to stop it before it happens.**

    If you have a close call, tell your teacher so that you and your teacher can find a way to prevent it from happening again. Otherwise, the next time, it could be a harmful accident instead of just a close call.

23. **All accidents should be reported to your teacher, no matter how minor.**

    Also, if you get a headache, feel sick to your stomach, or feel dizzy, tell your teacher immediately.

24. **For all chemicals, take only what you need.**

    On the other hand, if you do happen to take too much and have some left over, DO NOT put it back in the bottle. If somebody accidentally puts a chemical into the wrong bottle, the next person to use it will have a contaminated sample. Ask your teacher what to do with any leftover chemicals.

25. **NEVER take any chemicals out of the lab.**

    You should already know this rule.

26. **Horseplay and fooling around in the lab are very dangerous.**

    NEVER be a clown in the laboratory.

27. **Keep your work area clean and tidy.**

    After your work is done, clean your work area and all equipment.

28. **Always wash your hands with soap and water before you leave the lab.**

29. **Whether or not the lab instructions remind you, ALL of these rules APPLY ALL OF THE TIME.**

QUIZ

Determine which safety rules apply to the following.

- Tie back long hair, and confine loose clothing. (Rule ? applies.)
- Never reach across an open flame. (Rule ? applies.)
- Use proper procedures when lighting Bunsen burners. Turn off hot plates, Bunsen burners, and other heat sources when not in use. (Rule ? applies.)
- Heat flasks and beakers on a ring stand with wire gauze between the glass and the flame. (Rule ? applies.)
- Use tongs when heating containers. Never hold or touch containers with your hands while heating them. Always allow heated materials to cool before handling them. (Rule ? applies.)
- Turn off gas valves when not in use. (Rule ? applies.)

SAFETY SYMBOLS

To highlight specific types of precautions, the following symbols are used in the experiments. Remember that no matter what safety symbols and instructions appear in each experiment, all of the 29 safety rules described previously should be followed at all times.

Eye and clothing protection

- Wear laboratory aprons in the laboratory. Keep the apron strings tied so that they do not dangle.
- Wear safety goggles in the laboratory at all times. Know how to use the eyewash station.

Chemical safety

- Never taste, eat, or swallow any chemicals in the laboratory. Do not eat or drink any food from laboratory containers. Beakers are not cups, and evaporating dishes are not bowls.

- Never return unused chemicals to the original container.
- Some chemicals are harmful to the environment. You can help protect the environment by following the instructions for proper disposal.
- It helps to label the beakers and test tubes containing chemicals.
- Never transfer substances by sucking on a pipette or straw; use a suction bulb.
- Never place glassware, containers of chemicals, or anything else near the edges of a lab bench or table.

Caustic substances

- If a chemical gets on your skin or clothing or in your eyes, rinse it immediately, and alert your teacher.
- If a chemical is spilled on the floor or lab bench, tell your teacher, but do not clean it up yourself unless your teacher says it is OK to do so.

Heating safety

- When heating a chemical in a test tube, always point the open end of the test tube away from yourself and other people. (This is another new rule.)

Explosion precaution

- Use flammable liquids only in small amounts.
- When working with flammable liquids, be sure that no one else in the lab is using a lit Bunsen burner or plans to use one. Make sure there are no other heat sources present.

Hand safety

- Always wear gloves or cloths to protect your hands when cutting, fire polishing, or bending hot glass tubing. Keep cloths clear of any flames.

- Never force glass tubing into rubber tubing, rubber stoppers, or corks. To protect your hands, wear heavy leather gloves or wrap toweling around the glass and the tubing, stopper, or cork, and gently push in the glass tubing.
- Use tongs when heating test tubes. Never hold a test tube in your hand to heat it.
- Always allow hot glassware to cool before handling.

Glassware safety

- Check the condition of glassware before and after using it. Inform your teacher of any broken, chipped, or cracked glassware because it should not be used.
- Do not pick up broken glass with your bare hands. Place broken glass in a specially designated disposal container.

Gas precaution

- Do not inhale fumes directly. When instructed to smell a substance, use your hand, wave the fumes toward your nose, and inhale gently. (Some people say "waft the fumes.")

Radiation precaution

- Always wear gloves when handling a radioactive source.
- Always wear safety goggles when performing experiments with radioactive materials.
- Always wash your hands and arms thoroughly after working with radioactive materials.

Hygiene precaution

- Keep your hands away from your face and mouth.
- Always wash your hands before leaving the laboratory.

Any time you see any of the safety symbols you should remember that all 29 of the numbered laboratory rules always apply.

Labeling of Chemicals

In any science laboratory the *labeling* of chemical containers, reagent bottles, and equipment is essential for safe operations. Proper labeling can lower the potential for accidents that occur as a result of misuse. Labels and equipment instructions should be read several times before using. Be sure that you are using the correct items, that you know how to use them, and that you are aware of any hazards or precautions associated with their use.

All chemical containers and reagent bottles should be labeled prominently and accurately using labeling materials that are not affected chemicals.

Chemical labels should contain the following information.

1. **Name of chemical and the chemical formula**
2. **Statement of possible hazards** This is indicated by the use of an appropriate signal word, such as DANGER, WARNING, or CAUTION. This signal word usually is accompanied by a word that indicates the type of hazard present such as POISON, CAUSES BURNS, EXPLOSIVE or FLAMMABLE. Note that this labeling should not take the place of reading the appropriate Material Safety Data Sheet for a chemical.
3. **Precautionary measures** Precautionary measures describe how users can avoid injury from the hazards listed on the label. Examples include: "Use only with adequate ventilation," and "Do not get in eyes or on skin or clothing."
4. **Instructions in case of contact or exposure** If accidental contact or exposure does occur, immediate treatment is often necessary to minimize injury. Such treatment usually consists of proper first-aid measures that can be used before a physician administers treatment. An example is: "In case of contact, flush with large amounts of water; for eyes, rinse freely with water for 15 minutes and get medical attention immediately"
5. **The date of preparation and the name of the person who prepared the chemical** This information is important for maintaining a safe chemical inventory.

Suggested Labeling Scheme

Name of contents	Hydrochloric Acid	
	6 M HCl	Chemical formula and concentration or physical state
Statements of possible hazards and precautionary measures	WARNING! CAUSTIC and CORROSIVE-CAUSES BURNS CAUTION! Avoid contact with skin and eyes. Avoid breathing vapors.	
	IN CASE OF CONTACT: Immediately flush skin or eyes with large amounts of water for at least 15 minutes; for eyes, get medical attention immediately!	Hazard Instructions for contact or overexposure
Date prepared or obtained	May 8, 1989 Prepared by Betsy Byron Faribault High School, Faribault, Minnesota	Manufacturer (Commercially obtained) or preparer (Locally made)

Laboratory Techniques

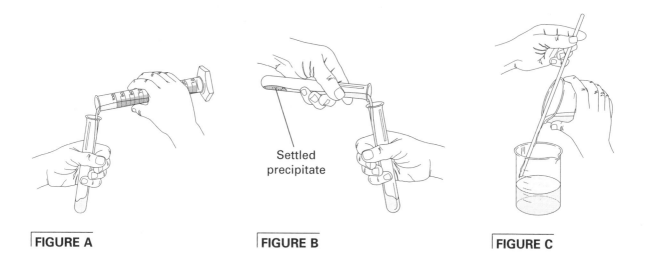

FIGURE A

Settled precipitate

FIGURE B

FIGURE C

DECANTING AND TRANSFERRING LIQUIDS

1. The safest way to transfer a liquid from a graduated cylinder to a test tube is shown in Figure A. The liquid is transferred at arm's length with the elbows slightly bent. This position enables you to see what you are doing and still maintain steady control.

2. Sometimes liquids contain particles of insoluble solids that sink to the bottom of a test tube or beaker. Use one of the methods shown below to separate a supernatant (the clear fluid) from insoluble solids.

a. Figure B shows the proper method of decanting a supernatant liquid in a test tube.

b. Figure C shows the proper method of decanting a supernatant liquid in a beaker by using a stirring rod. The rod should touch the wall of the receiving container. Hold the stirring rod against the lip of the beaker containing the supernatant liquid. As you pour, the liquid will run down the rod and fall into the beaker resting below. Using this method, the liquid will not run down the side of the beaker from which you are pouring.

HEATING SUBSTANCES AND EVAPORATING SOLUTIONS

1. Use care in selecting glassware for high-temperature heating. The glassware should be heat resistant.

2. When heating glassware using a gas flame, use a ceramic-centered wire gauze to protect glassware from direct contact with the flame. Wire gauzes can withstand extremely high temperatures and will help prevent glassware from breaking. Figure D shows the proper setup for evaporating a solution over a water bath.

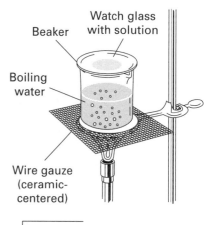

Watch glass with solution
Beaker
Boiling water
Wire gauze (ceramic-centered)

FIGURE D

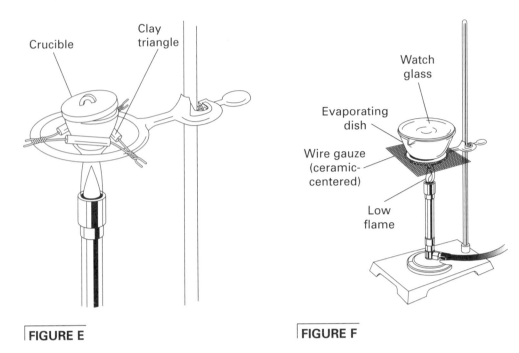

FIGURE E

FIGURE F

3. In some experiments you are required to heat a substance to high temperatures in a porcelain crucible. Figure E shows the proper apparatus setup used to accomplish this task.

4. Figure F shows the proper setup for evaporating a solution in a porcelain evaporating dish with a watch glass cover that prevents spattering.

5. Glassware, porcelain, and iron rings that have been heated may *look* cool after they are removed from a heat source, but these items can still burn your skin even after several minutes of cooling. Use tongs, test-tube holders, or heat-resistant mitts and pads whenever you handle this apparatus.

6. You can test the temperature of questionable beakers, ring stands, wire gauzes, or other pieces of apparatus that have been heated by holding the back of your hand close to their surfaces before grasping them. You will be able to feel any heat generated from the hot surfaces. DO NOT TOUCH THE APPARATUS. Allow plenty of time for the apparatus to cool before handling.

HOW TO POUR LIQUID FROM A REAGENT BOTTLE

1. Read the label at least three times before using the contents of a reagent bottle.

2. Never lay the stopper of a reagent bottle on the lab table.

3. When pouring a caustic or corrosive liquid into a beaker use a stirring rod to avoid drips and spills. Hold the stirring rod against the lip of the reagent bottle. Estimate the amount of liquid you need and pour this amount along the rod into the beaker See Figure G.

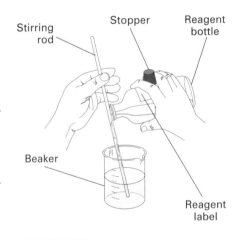

FIGURE G

ChemFile LAB D

4. Extra precaution should be taken when handling a bottle of acid. Remember the following important rules: Never add water to any concentrated acid, particularly sulfuric acid, because the mixture can splash and will generate a lot of heat. To dilute any acid, add the acid to water in small quantities, while stirring slowly. Remember the "triple A's"-Always Add Acid to water.

5. Examine the outside of the reagent bottle for any liquid that has dripped down the bottle or spilled on the counter top. Your teacher will show you the proper procedures for cleaning up a chemical spill.

6. Never pour reagents back into stock bottles. At the end of the experiment, your teacher will tell you how to dispose of any excess chemicals.

HOW TO HEAT MATERIAL IN A TEST TUBE

1. Check to see that the test tube is heat-resistant.

2. Always use a test tube holder or clamp when heating a test tube.

3. Never point a heated test tube at anyone, because the liquid may splash out of the test tube.

4. Never look down into the test tube while heating it.

5. Heat the test tube from the upper portions of the tube downward and continuously move the test tube as shown in Figure H. Do not heat any one spot on the test tube. Otherwise a pressure build-up may cause the bottom of the tube to blow out.

HOW TO USE A MORTAR AND PESTLE

1. A mortar and pestle should be used for grinding only one substance at a time. See Figure I.

2. Never use a mortar and pestal for simultaneously mixing different substances.

3. Place the substance to be broken up into the mortar

4. Pound the substance with the pestle and grind to pulverize.

5. Remove the powdered substance with a porcelain spoon.

DETECTING ODORS SAFELY

1. Test for the odor of gases by wafting your hand over the test tube and cautiously sniffing the fumes as shown in Figure J.

2. Do not inhale any fumes directly.

3. Use a fume hood whenever poisonous or irritating fumes are evolved. DO NOT waft and sniff poisonous or irritating fumes.

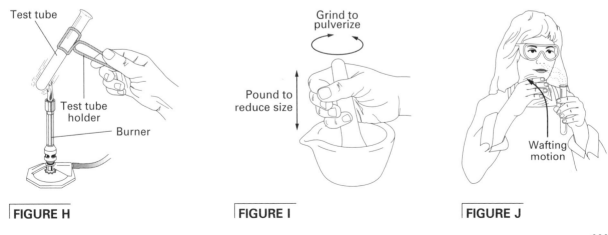

FIGURE H   **FIGURE I**   **FIGURE J**

ChemFile LAB D   **xiii**

Name_____

Date_____ Class_____

EXPERIMENT

How Sweet It Is!

OBJECTIVES

Recommended time:
45 minutes

- **Test** for the presence of sugars in beverages.
- **Determine** which is denser—a diet beverage or a regular beverage.

INTRODUCTION

Solution/ Material Preparation

1. To prepare 1 L of Benedict's solution (quantitative), dissolve 200 g of crystalline sodium carbonate, 200 g of sodium citrate, and 125 g of potassium thiocyanate in about 800 mL of distilled water. Use heat, and filter if necessary. Now dissolve 18 g of crystallized $CuSO_4$ in about 100 mL of distilled water, and slowly pour this solution into the citrate-carbonate-thiocyanate solution while stirring continuously. Dilute this solution to exactly 1 L. **NOTE:** 25 mL of this reagent is reduced by 50 mg of glucose.

Sucrose, or table sugar, was considered the universal beverage sweetener until the early 1960's, when bottlers created a new variety of diet beverages that contained artificial sweeteners such as cyclamates and saccharin. Cyclamates were banned as a food additive in the early 1970s, due to concern over their safety. Saccharin was first compounded in 1879 and was used mainly as a sugar-free sweetener in diabetic foods and later in beverages. By 1977, Americans were consuming 4.5 million pounds of saccharin in beverages yearly! Today saccharin stands in the shadow of another artificial sweetener, aspartame, which is a combination of two amino acids: aspartic acid and l-phenylalanine.

Bottlers now use high fructose corn syrup as their primary sugar sweetener. Cornstarch is the raw material used in the production of corn syrup. Glucose isomerase, an enzyme, is used to convert corn syrup into a mixture of glucose and fructose. High glucose fractions are used in the food industry as thickeners, while high fructose fractions are used as sweeteners in soft drinks.

Aspartame, the newest artificial sweetener, provides about 4 Calories per gram and is 160 times as sweet as sucrose. Aspartame provides the equivalent sweetness of table sugar with about 1/160 as many calories. Because it takes less artificial sweetener than sugar to achieve the same degree of sweetness, diet drinks are less dense. Therefore, density can be used to distinguish between a diet beverage and a regular beverage. In this experiment, you will test colorless beverages for the presence of sugar and determine their densities. Benedict's solution tests for the presence of simple (reducing) sugars such as the monosaccharide glucose. Sucrose is a disaccharide formed from glucose and fructose.

SAFETY

2. Prepare an ingredient list for each beverage tested. Following **Part 2,** pass out copies of the list to students to check their analysis results. A typical national brand carbonated beverage formula (ingredients listed from highest to lowest percent) is:

Always wear safety goggles and a lab apron to protect your eyes and clothing. If you get a chemical in your eyes, immediately flush the chemical out at the eyewash station while calling to your teacher. Know the location of the emergency lab shower and the eyewash station and the procedures for using them.

Do not touch any chemicals. If you get a chemical on your skin or clothing, wash the chemical off at the sink while calling to your teacher. Make sure you carefully read the labels and follow the precautions on all containers of chemicals that you use. If there are no precautions stated on the

carbonated water, high fructose corn syrup and/or sugar, natural flavorings, citric acid and sodium citrate (flavor enhancers), sodium benzoate (preservative), EDTA (preservative), caffeine, gum arabic and bromated vegetable oil (provide body), colorings.

label, ask your teacher what precautions you should follow. Do not taste any chemicals or items used in the laboratory. Never return leftovers to their original containers; take only small amounts to avoid wasting supplies.

Call your teacher in the event of a spill. Spills should be cleaned up promptly, according to your teacher's directions.

Never put broken glass in a regular waste container. Broken glass should be disposed of properly.

MATERIALS

Required Precautions

• Read all safety precautions, and discuss them with your students.
• Safety goggles and an apron must be worn at all times.

- 15 mL Benedict's solution
- 5 brands of clear soft-drink beverages
- centigram balance
- 500 mL beaker
- 100 mL graduated cylinder
- 10 mL graduated cylinder
- hot plate

- iron ring large enough to fit around the 500 mL beaker
- ring stand
- test tubes, 5
- test-tube clamp
- test-tube rack
- wax pencil

PROCEDURE

• In case of a spill, use a dampened cloth or paper towels to mop up the spill. Then rinse the cloth in running water at the sink, wring it out until it is only damp, and put it in the trash.
• Broken glass should be disposed of in a clearly labeled box lined with a plastic trash bag. When full, close the box, seal it with packaging tape, and set it next to the trash can for disposal.

Part 1: Testing for the presence of simple sugars in beverages

1. Prepare a water bath by filling a 500 mL beaker to the 200 mL mark with tap water. Set the beaker on a hot plate. Attach an iron ring to a ring stand. Place the ring stand next to the hot plate and lower the ring until it is 2 cm below the rim of the beaker. While the water is warming to a slow boil, complete steps 2 through 4.

2. Use a wax pencil to label five test tubes 1 to 5.

3. Use a graduated cylinder to measure 5 mL of a beverage sample. Pour the sample into its corresponding test tube. Set the test tube in a test-tube rack. Do the same for the remaining 4 beverage samples.

4. Use a graduated cylinder to measure 3 mL of Benedict's solution. Pour the solution into test tube 1. Record the color of the resultant mixture in **Data Table 1.** Do the same for test tubes 2 through 5.

5. Use a test-tube clamp to handle the test tubes. Carefully place a test tube containing the mixture of beverage sample and Benedict's solution in the water bath for at least 5 minutes.

6. Using a test-tube clamp, remove the test tube from the water bath and place it in the test-tube rack. Observe the mixture's color and its intensity (light or dark). Record these values in **Data Table 1.** A color change to red, orange, or yellow following heating indicates the presence of reducing sugars. The darker the color is, the greater the amount of sugar present. A precipitate may also be present in a positive test.

Procedural Tips

• Select beverage samples from the following categories: carbonated water; a clear, carbonated regular and diet beverage of the same brand; tonic water; clear, uncarbonated sports drink such as Gatorade©.

• High fructose corn syrup tests positive with Benedict's solution due to the presence of glucose. Beverages containing natural fruit juices test negative because fructose is not a reducing sugar.

• If available, pipets offer a greater accuracy when dispensing the 100 mL volumes in steps 9 and 11.

Techniques to Demonstrate

• Review the proper procedure for reading a meniscus.

Pre-Lab Discussion

• Discuss with students how to identify a positive and a negative Benedict's result. You may want to tell students that other tests, such as the Seliwanoff, are used to detect nonreducing sugars like the fruit sugar, fructose.

• Caution students that experimental results depend on both the accuracy with which masses are obtained and with which the volumes are measured.

• Review density calculations, and remind students to record all digits shown on the balance.

Data Table 1

Sample number	Color of unheated sample	Color of heated sample	Color intensity (light/dark)	Sugar present (√)
1	Blue			
2	Blue	Answers will vary depending upon		
3	Blue	the brand selected.		
4	Blue			
5	Blue			

7. Repeat steps 5 and 6 for each sample.

Part 2: Determining density of beverages

8. Measure the mass of a clean, dry 100 mL graduated cylinder. Record the value in **Data Table 2**.

9. Add 100 mL of distilled water to the graduated cylinder. Determine the mass of the cylinder and distilled water. Record this value in **Data Table 2**. Water is used in this experiment as a reference standard. Its density is 1.00 g/mL. In terms of the scientific method, is water considered a control or a variable?

a control

10. Rinse and dry the graduated cylinder. Remeasure its mass and record the value in **Data Table 2**.

11. Add 100 mL of a beverage sample to the graduated cylinder. Determine the mass of the cylinder and the sample. Record this value in **Data Table 2**.

12. Repeat steps 10 and 11 for each of the remaining beverage samples and record their respective masses in **Data Table 2**. Complete the Calculations column.

13. Using the Information Table on the next page and your results from **Data Table 1** and **Data Table 2**, record your conclusions about the identity of each sample in **Data Table 3**.

Cleanup and Disposal

14. Clean all apparatus and your lab station. Return equipment to its proper place. Dispose of chemicals and solutions in the containers designated by your teacher. Do not pour any chemicals down the drain or put them in the trash unless your teacher directs you to do so. Wash your hands thoroughly after all work is finished and before you leave the lab.

DATA TABLE 1:
● Sports drinks contain sucrose, which is not a reducing sugar, but they also contain some glucose and will therefore test positive with Benedict's. Test reactions will vary based upon the amount of reducing sugars that contain an aldehyde group, which is the basis for this type of reaction.

DATA TABLE 2:
Diet beverages will have a negative Benedict's test, and their densities will be less than the densities of the corresponding regular brand. Tonic water tests as regular beverage. Carbonated water has a density close to 1.00 g/mL unless small amounts of mineral salts are present.

Data Table 2

Sample number	Mass of empty cylinder	Mass of cylinder + sample	Calculated density value (g/mL)
Water			1.00 g/mL
1			
2	Answers will vary depending on the beverages chosen.		
3			
4			
5			

Information Table

Sweetner	Density (g/mL)
Glucose	1.0
Fructose	1.7
Sucrose	1.6
Saccharin	0.8
Aspartame	0.9

Data Table 3

Sample number	Experimental conclusion* [√]	Manufacturer's sweetner ingredient(s)
1	Regular [ ] Diet [ ] Other [ ]	Answers will
2	Regular [ ] Diet [ ] Other [ ]	vary depending on the beverages chosen.
3	Regular [ ] Diet [ ] Other [ ]	
4	Regular [ ] Diet [ ] Other [ ]	
5	Regular [ ] Diet [ ] Other [ ]	

* *Regular* indicates that the product contains sugar; *Diet* indicates that the product does not contain sugar; *Other* implies that the product contains no sweetener.

CALCULATIONS

1. **Organizing Data** Calculate the mass of each sample tested by subtracting the mass of the empty graduated cylinder from the mass of the cylinder plus the liquid sample.

   Example data: Combined mass = 510 g

   $$
   \begin{array}{ll}
   - \text{ Cylinder mass} & = 350 \text{ g} \\
   \hline
   \text{Liquid mass} & = 160 \text{ g}
   \end{array}
   $$

2. **Organizing Data** Calculate the density of each sample tested by dividing each mass by 100 mL, the volume of each sample. Record your values in **Data Table 2.**

   Example data: $\dfrac{160 \text{ g}}{100 \text{ mL}}$ (sample volume) = 1.6 g/mL

QUESTIONS

1. **Analyzing Data** Review the ingredients list for each sample provided by your teacher. Record and compare the "Manufacturer's sweetener ingredient(s)" for each sample to your results. Record the sweetener(s) in the corresponding area of **Data Table 3.** How accurate was your analysis?

   Answers will vary. Beverages that contain a sugar sweetener

   should be identifiable to students based on a positive Benedict's

   test. There should be some general correlation between the

   sugar test and the density calculation. A negative Benedict's

   test and a low density calculation (near 1.0) should indicate the

   presence of an artificial sweetener—a diet beverage. A control

   sample of carbonated water should have a density calculation of

   1 g/mL and a negative Benedict's test and should be classified as

   other.

2. Analyzing Data Which sample had the highest sugar content, and which had the lowest?

Answers will vary. The color intensity in the Benedict's test is re-

lated to the amount of sugar present. About 25 mL of this quan-

titative reagent is reduced by 50 mg of glucose.

GENERAL CONCLUSIONS

1. Applying Concepts and Designing Experiment Explain how you could prove that carbonated water contains no sweetener.

Measure and find the mass of 100 mL of distilled water. Carbon-

ated water contains no sweeteners. The density of exactly

100 mL of this beverage will be 1, the density of water. A Bene-

dict's test will be negative.

2. Designing Experiments Examine the densities you recorded in **Data Table 2.** Use this information to design an experiment that distinguishes between the density of a regular soft drink and that of a diet soft drink without opening the beverage containers.

Obtain two unopened cans of the same soft drink, one with

sugar and the other with sweetener. The samples must have the

same volume. Immerse each can into a bucket of water so it just

floats lengthwise on the surface. Gently tap each with a finger.

The can with the *regular* soft drink will sink, while the *diet* soft

drink floats.

Name_____

Date_____ Class_____

What's So Special About Bottled Drinking Water?

OBJECTIVES

Recommended time:
45 minutes

INTRODUCTION

Required Precautions

● Read all safety precautions, and discuss them with your students.
● Safety goggles and an apron must be worn at all times.
● In case of a spill, use a dampened cloth or paper towels to mop up the spill. Then rinse the cloth in running water at the sink, wring it out until it is only damp, and put it in the trash.

Techniques to Demonstrate

● Demonstrate how to determine when to stop adding water to the polymer.

SAFETY

Solution/Material Preparation

1. In a common work area, set out bottled water from at least five separate sources. Include distilled water, mineral water, and imported and domestic drinking water.
2. If a commercial source of sodium polyacrylate is unavailable, remove some from disposable diapers.

● **Conduct** an audit of various brands of bottled water.

● **Compare** the mineral content of various brands of bottled waters.

The International Bottled Water Association (IBWA) is the trade association that represents the bottled-water industry. Founded in 1958, IBWA's member companies produce and distribute more than 85% of the bottled water sold in the United States. Within the United States, bottled water is regulated as a food by the U.S. Food and Drug Administration (FDA). In contrast, municipal water is regulated as a commodity by the Environmental Protection Agency (EPA).

The FDA has labeling rules and regulations for bottled water. A table of definitions for bottled water labels appears on the next page.

Labels must include the name of the manufacturer, a statement of the "net contents," and a list of ingredients if more than one ingredient is present. Any nutrient content claims must meet additional FDA regulations. "Sodium free" is an example of a nutrient content claim. When errors occur in the labeling of a product, or when a label intentionally misrepresents the product, the product has been *misbranded*. To correct a misbranding, the FDA works with the manufacturer through voluntary compliance, civil action, or criminal action, depending on the circumstances.

In this experiment, you will examine the labels of bottled-water samples and determine the amount of each sample that is absorbed by 0.1 g of sodium polyacrylate. The amount of water absorbed is related to the amount of mineral salts in the water.

  Always wear safety goggles and a lab apron to protect your eyes and clothing. If you get a chemical in your eyes, immediately flush the chemical out at the eyewash station while calling to your teacher. Know the location of the emergency lab shower and the eyewash station and the procedures for using them.

 Do not touch any chemicals. If you get a chemical on your skin or clothing, wash the chemical off at the sink while calling to your teacher. Make sure you carefully read the labels and follow the precautions on all containers of chemicals that you use. If there are no precautions stated on the label, ask your teacher what precautions you should follow. Do not taste any chemicals or items used in the laboratory. Never return leftovers to their original containers; take only small amounts to avoid wasting supplies.

Procedural Tips

Pool class data for a more accurate graph on which to plot unknown values.

Call your teacher in the event of a spill. Spills should be cleaned up promptly, according to your teacher's directions.

Never put broken glass in a regular waste container. Broken glass should be disposed of properly.

MATERIALS

If materials are available, have students take multiple measurements for each sample and graph the average of the results.

- sodium polyacrylate
- balance
- 7 oz plastic cups
- 100 mL graduated cylinder
- microspoon or microspatula
- micropipet or medicine dropper
- wax pencil
- weighing boat

Standardized Definitions for Types of Bottled Water

Artesian water/ artesian well water	Bottled water from a well that taps a confined aquifer (a water-bearing underground layer of rock or sand) in which the water stands at some height from the top of the aquifer.
Drinking water	Bottled water sold for human consumption, containing no additives other than flavors, extracts, or essences less than 1% by mass; calorie free and sugar free.
Mineral water	Bottled water containing not less than 250 mg/L (ppm) total dissolved solids; has a constant level and regular proportion of mineral and trace element composition at the point of emergence from the source; no minerals can be added
Purified water	Water produced by distillation, deionization, reverse osmosis, ozonation, or other suitable process; other suitable names include distilled water or reverse osmosis water.
Spring water	Bottled water derived from an underground formation, that flows naturally to the surface; spring water must be collected at the spring or through a bore hole tapping the underground formation
Sparkling water	Water that contains the same amount of carbon dioxide that it had at emergence from the source. This label does not pertain to seltzers, soda, or tonic water, which are soft drink beverages.
Well water	Bottled water from a well bored or drilled in the ground, which taps the water of an aquifer.

ChemFile

PROCEDURE

Pre-Lab Discussion

● When recording marketing claims, students should pay careful attention to the use of nonregulated terminology such as purity guaranteed; crisp, refreshing taste; and graphics that imply a specific source, such as a mountain stream or spring.
● You might point out to students that labels do not use the word *natural* as a standard of identity because the FDA has not defined *natural*.
● Point out to students that seltzers, soda water, and tonic water are considered beverages, not bottled water.
● The amount of dissolved solid is reported either in parts per million (ppm) or in milligrams per liter (mg/L). These units are equivalent.
● Most imported brands of bottled water and some domestic brands of bottled water are naturally carbonated because carbon dioxide gas present underground is dissolved in the water. Mineral salts help retain carbon dioxide. They bind with CO_2 resulting in water that has smaller bubbles. Most imported mineral water has very small gas bubbles. Domestic sparkling water also contains gas bubbles, but these may be larger and more numerous because of the recarbonation following purification.

Part 1: Conduct a bottled water audit

1. Obtain five samples of bottled water from your teacher. Write the name of each sample in the **Data Table.**

2. Carefully read the labels on each sample product. Then enter the data requested in the **Data Table.** Use the table of standardized definitions for types of bottled water to identify the product label type.

Part 2: Compare the mineral content of the bottled water samples

3. Use a wax pencil to label a set of five transparent plastic cups 1 to 5.

4. Add 0.1 g sodium polyacrylate to each cup.

5. Carefully open each sample bottle. Note the size and number of gas bubbles formed, if any. Record your findings in the **Data Table.**

6. Fill a graduated cylinder to the 100 mL mark with water from sample bottle 1.

7. Using a micropipet or medicine dropper, remove water from the graduated cylinder and add it dropwise to the polymer in cup 1. Keep adding drops until the white powder is no longer visible. *You may need to carefully observe the underside of the cup to make sure.*

8. When the white powder has dissolved, empty any water remaining in the micropipet or medicine dropper into the graduated cylinder. Record the volume used in the **Data Table.**

9. Repeat steps 6, 7, and 8 for each of the remaining water samples.

10. On the grid provided, graph "Volume of water absorbed (mL)" versus "Amount of dissolved solids (ppm)" for each sample tested. Distilled water has 0.0 dissolved solids. Draw a line of best fit through data points. Use this graph to estimate the amount of dissolved solids for the samples that do not indicate a value for dissolved solids.

Cleanup and Disposal

11. Clean all apparatus and your lab station. Return equipment to its proper place. Dispose of chemicals and solutions in the containers designated by your teacher. Do not pour any chemicals down the drain or put them in the trash unless your teacher directs you to do so. Wash your hands thoroughly after all work is finished and before you leave the lab.

Data Table

	Sample 1 Aquafina	Sample 2 Distilled Water	Sample 3 Gerolsteiner	Sample 4 Evian	Sample 5 Perrier
Product label description	Purified drinking water	Purified water	Sparkling mineral water	Spring water water	Sparkling mineral water
Carbonated	Yes [ ] No [√]	Yes [ ] No [√]	Yes [√] No [ ]	Yes [ ] No [√]	Yes [√] No [ ]
Method of purification	Reverse osmosis	Distillation, ozonation, carbon filtered	None	None	None
Dissolved solids (mg/L or ppm)	Not indicated	0.0	2527	309	475
Product packaging	Glass [ ] Plastic [√] Color [ ] No color [√]	Glass [ ] Plastic [√] Color [ ] No color [√]	Glass [√] Plastic [ ] Color [ ] No color [√]	Glass [ ] Plastic [√] Color [ ] No color [√]	Glass [√] Plastic [ ] Color [√] No color [ ]
Collection source on label	Yes [ ] No [√]	Yes [ ] No [√]	Yes [√] No [ ]	Yes [√] No [ ]	Yes [√] No [ ]
Marketing claims	Purity guaranteed; crisp refreshing taste	None	Low sodium; contains natural calcium; naturally sparkling	As pure and natural as when its source was discovered	Low mineral content
Size of carbonation bubbles	N/A	N/A	Tiny	N/A	Small
Volume of water absorbed (mL)	52	65	3	42	25

Aquafina had an absorbency of approximately 52 mL. Its dissolved solids content is approximately 200 mg/L.

● Most brands do not identify a collection source on the label except for "Bottled at the source."

● All domestic national, regional, and generic brands will list some form of purification; some brands will list multiple types. Imported brands do not list any purification measures, they simply acknowledge that the source is protected. The European community requires mineral water to be free of specified chemical and microbiological impurities; it must be bottled at the source and cannot be filtered or treated in any way.

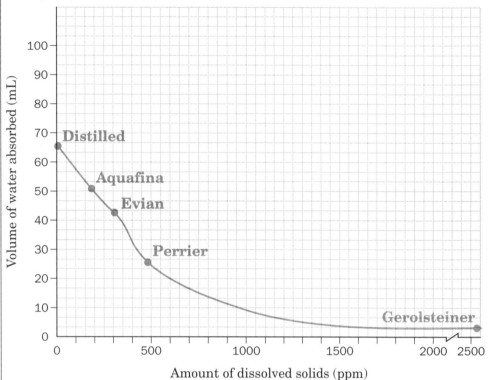

Absorbency versus Mineral Concentration

QUESTIONS

Disposal

Water samples may be poured down the drain. Used polymer samples may be placed in the trash. Bottles should be placed in appropriate recycling containers.

1. **Analyzing Data** How is the volume of water absorbed, known as absorbency, related to the amount of dissolved solids?

Absorbency decreases as the dissolved solid concentration (mineral content) increases.

2. **Inferring Conclusions** If your data were used to determine compliance with labeling regulations, would any of the water samples tested be misbranded? Justify your answer.

Answers will vary. Examples of misbranding are as follows: If high levels of mineral salts are present, more than a single ingredient is present, and there should be a list of ingredients on the product label. If no list were present, the product would be misbranded. If a sample is labeled as sodium-free but has a low polymer absorbency level, further testing may be needed.

GENERAL CONCLUSIONS

1. **Inferring Conclusions** Spring water tends to be the predominant type of domestic and regional bottled water. Very few domestic brands are labeled mineral water; most bottled mineral water is imported. Suggest a reason why domestic companies do not market mineral water even though FDA regulations permit them to.

Mineral water lacks market appeal. Apparently, Americans prefer

spring water, perhaps because the term sounds fresh and pure.

Accept all reasonable answers citing economic or marketing

issues.

2. **Analyzing Data and Inferring Conclusions** Examine your packaging data. Which type of water comes in colored containers? Suggest a reason for the color. (Hint: Examine the method-of-purification data.)

Mineral water comes in colored containers. The color protects

the water from sunlight, which causes algal growth.

Name_____

Date_____ Class_____

A Close Look at Toothpaste

OBJECTIVES

Recommended time:
45 minutes

- **Compile** a scratch index of the abrasive materials used in a toothpaste.
- **Test** for the presence of carbonate material in toothpaste.
- **Compare** the pH values of different brands of toothpaste to the pH value of saliva.

INTRODUCTION

Solution/Material Preparation

1. Wear safety goggles, a face shield, impermeable gloves, and an apron when you prepare the HCl solution. Work in a chemical fume hood known to be in operating condition and have another person stand by to call for help in case of an emergency. Be sure you are within a 30 second walk from a safety shower and eyewash station known to be in good operating condition.

A healthy tooth is protected by a coating of enamel. Enamel is one of the hardest biomaterials known. It is composed of a calcium-rich mineral salt matrix termed hydroxyapatite, $Ca_{10}(PO_4)_6(OH)_2$. Despite its toughness, enamel can be broken down by acids and enzymes in plaque, which is a thin, adhesive, polysaccharide film. Bacteria in the mouth are constantly converting the sugars in foods into plaque. Other bacteria convert plaque into an acid that erodes the calcium and phosphate shield of hydroxyapatite. When enough erosion of the enamel occurs, bacteria can work their way into the tooth's interior. The result is tooth decay.

The key to healthy teeth is the removal of plaque. Calcium ions and phosphate ions normally present in saliva can replace ions that are lost when plaque is removed. Remineralization is part of a reversible reaction that can return enamel to its original strength.

$$\underset{\text{Hydroxyapatite}}{Ca_5(PO_4)_3OH} \underset{\text{remineralization}}{\overset{\text{demineralization}}{\rightleftharpoons}} 5Ca^{2+} + 3PO_4^{3-} + OH^-$$

To be effective, a toothpaste, or dentifrice, must contain an abrasive material that removes plaque but does not remove tooth enamel.

SAFETY

2. To prepare 1 000 mL of 1.0 M HCl solution, observe the required safety precautions. Slowly add 86 mL of 12 M HCl to 500 mL of distilled water while stirring continuously. Dilute the solution to 1000 mL.

Always wear safety goggles and a lab apron to protect your eyes and clothing. If you get a chemical in your eyes, immediately flush the chemical out at the eyewash station while calling to your teacher. Know the location of the emergency lab shower and the eyewash station and the procedures for using them.

Do not touch any chemicals. If you get a chemical on your skin or clothing, wash the chemical off at the sink while calling to your teacher. Make sure you carefully read the labels and follow the precautions on all containers of chemicals that you use. If there are no precautions stated on the label, ask your teacher what precautions you should follow. Do not taste any chemicals or items used in the laboratory. Never return leftovers to their original containers; take only small amounts to avoid wasting supplies.

3. To prepare limewater solution, observe the required safety precautions. Avoid dusting. Add 700 mg Ca(OH)$_2$ to 500 mL of distilled water. **NOTE:** An excess of Ca(OH)$_2$ should be added to assure a saturated solution. Seal the container, carefully shake, and let stand for 24 hours. Then pour off the supernatant fluid, filter the solution if necessary, and keep it well-sealed.

Acids and bases are corrosive. If an acid or base spills onto your skin or clothing, wash the area immediately with running water. Call your teacher in the event of an acid spill. Acid or base spills should be cleaned up promptly.

 Call your teacher in the event of a spill. Spills should be cleaned up promptly, according to your teacher's directions.

Never put broken glass in a regular waste container. Broken glass should be disposed of properly.

 Wear disposable plastic gloves to handle the microscope slides.

MATERIALS

4. To make the rubber-stopper assemblies, insert a 3.0 cm length of 4 mm glass or plastic tubing into a one-holed No. 2 stopper. To prevent cuts in case the glass tubing breaks, hold the tubing with a cloth while inserting it into the stopper.

- 5 brands of toothpaste
- 1.0 M HCl, 30 mL
- limewater solution
- distilled water
- 2 oz paper cups, 5
- 250 mL beaker
- 100 mL graduated cylinder
- 10 test tubes, 20 mm × 150 mm
- test-tube clamp
- 4 mm glass tubing, 3.0 cm

- rubber stopper, one-hole, No. 2
- 1/8 in. rubber tubing, 12 in.
- plastic microscope slides, 10
- lens tissue
- compound microscope or magnifying glass
- flat toothpicks, 5
- universal pH paper
- wax pencil

PROCEDURE

5. Select five toothpaste brands (gels, drops, and pastes). Label the brands 1 to 5, and record the ingredients for each brand. Make copies of these ingredients lists for students to use in Step 10. **NOTE:** Manufacturers provide product contents in order of decreasing volume. A typical dentifrice formula is: Water (37%), Glycerol-humectant, retains moisture (32%), Calcium carbonate-abrasive (27%), Sodium N-lauryl sarcosinate-surfactant (foaming) (2%), Carrageenan-thickening agent (1%), Fluorides and other additives, enamel hardener and preservatives (1%).

Part 1: Compiling a Scratch Index of the abrasive material used in toothpaste

1. Identify the abrasive material(s) in the list of ingredients for each toothpaste sample tested. Look for materials such as carbonates, silicas, and oxides. Record the brand name, toothpaste type (gel, paste, or drop), and material name in **Data Table 1.**

2. Use a wax pencil to label five plastic slides 1 to 5. Use a toothpick to place a small amount of each numbered toothpaste sample in the middle of the corresponding plastic slide.

3. Place a clean plastic slide on slide 1. Gently rub the two slides together 50 times as illustrated in **Figure A.** Repeat for slides 2 through 5.

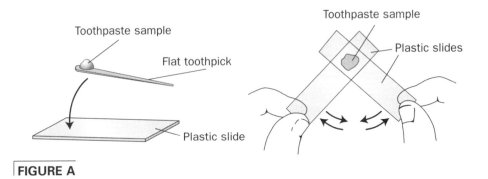

FIGURE A

4. Remove the labeled slide, wash it under running tap water, and dry it with clean lens tissue. Examine each labeled slide under low power (40×) of a compound microscope (or a 10× magnifying glass) for scratches.

5. In **Data Table 1,** record the abrasive action of each toothpaste sample as a numerical value. Use 1–3 points for "Light," 4–6 points for "Moderate," and 7–9 points for "Heavy," depending upon the depth and number of scratch markings on the plastic plate. Use scratch index data to determine which abrasive material(s) is most effective.

Data Table 1

Sample number	Scratch index abrasive rating	Toothpaste type	Manufacturer's brand name	Manufacturer's abrasive ingredient
1				
2				
3				
4				
5				

Most effective abrasive material: _Answers will vary._

Required Precautions

● Read all safety precautions, and discuss them with your students.
● Safety goggles and an apron must be worn at all times.
● In case of a spill, use a dampened cloth or paper towels to mop up the spill. Then rinse the cloth in running water at the sink, wring it out until it is only damp, and put it in the trash.
● Broken glass should be disposed of in a clearly labeled box lined with a plastic trash bag. When the box is full, close it, seal it with packaging tape, and set it next to the trash can for disposal.
● Students should avoid direct contact with the HCl solution. If contact occurs, the affected areas should be thoroughly rinsed with water.

Part 2: Testing for carbonates

6. Use a wax pencil to label five test tubes 1 to 5. Referring to **Figure B,** attach a small length of rubber tubing to a rubber stopper fitted with a short piece of glass tubing.

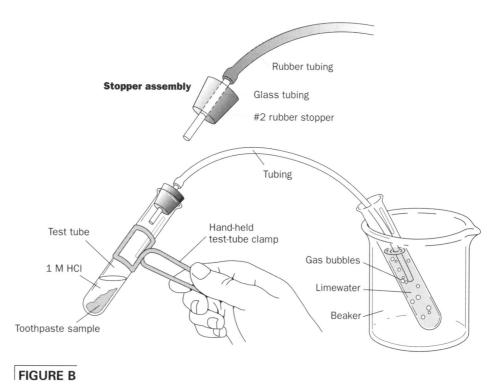

Stopper assembly

Rubber tubing

Glass tubing

#2 rubber stopper

Tubing

Test tube

Hand-held test-tube clamp

1 M HCl

Gas bubbles

Limewater

Beaker

Toothpaste sample

FIGURE B

Materials

● **NOTE:** The 4 mm glass tubing is an outer diameter measure. The 1/8 in. rubber tubing is an inner diameter measure.

Techniques to Demonstrate

● Demonstrate the slide preparation illustrated in **Figure A** of the pupil text and the use of a compound microscope.

Pre-Lab Discussion

● Tell students that the pH value of saliva is 6.8.
● Students should understand what is meant by pH reading and that the lower the pH value, the more acidic the tested substance.
● Discuss with students the interrelationship of the demineralization and remineralization processes. They should realize that in acidic environments hydroxyapatite dissociates into ions, while in neutral and basic environments hydroxyapatite is produced.
● You may want to review crystal lattice formation.

7. Fill an unmarked test tube with 15 mL of limewater (calcium hydroxide solution). Obtain about 1 g of a sample toothpaste. Place it in a numbered test tube so that it fills a 1 cm depth. Attach the test-tube clamp to the test tube, and add 6 mL of 1 M HCl to the sample in the test tube. Seal the numbered test tube with the rubber stopper assembly, and place the free end of the rubber tubing into the test tube containing limewater.

8. Record in **Data Table 2** whether a gas evolves, and if the test for carbonate is positive (limewater solution turns cloudy). What gas is released? Why do you think limewater turns cloudy?

CO_2 gas is released when carbonate is present. Brands containing

sodium carbonate peroxide also release H_2O_2 gas. Limewater

turns cloudy when $CaCO_3$ precipitates.

_____

9. Repeat steps 7 and 8 for each sample tested. Which sample(s) contain carbonates?

All samples that have a positive limewater test contain carbonates.

_____

10. Compare your carbonate analysis with the list of ingredients for each sample. Enter the ingredient name under "Manufacturer's ingredient" in **Data Table 2.** How accurate was your analysis?

Accept all reasonable answers.

_____

_____

Data Table 2

Sample number	Gas evolved? ($\sqrt{}$)	Limewater turned cloudy? ($\sqrt{}$)	Positive for carbonate? ($\sqrt{}$)	Manufacturer's ingredient	pH
1					
2					
3					
4					
5					

Procedural Tips

• Generally, non-carbonate abrasives such as hydrated silica have a higher Scratch Index. Some compounds used by manufacturers are: calcium carbonate, calcium pyrophosphate, dibasic calcium phosphate, hydrated silica, hydrated aluminum oxide, magnesium carbonate, talc, titanium dioxide.

Part 3: Comparing pH values of toothpaste and saliva

11. Use a wax pencil to label five cups 1 to 5. Use a toothpick to place a dab of toothpaste sample 1 in the corresponding cup. Add 10 mL of distilled water to the cup, and stir to mix. Dip a small piece of universal pH paper into the toothpaste solution. Record the pH value in **Data Table 2.** Repeat with toothpaste samples 2–5. Which brand has a pH value closest to that of human saliva (pH = 6.8)?

Answers will vary according to toothpaste brands selected.

Cleanup and Disposal

12. Clean all apparatus and your lab station. Return equipment to its proper place. Dispose of chemicals and solutions in the containers designated by your teacher. Do not pour any chemicals down the drain or put them in the trash unless your teacher directs you to do so. Wash your hands thoroughly after all work is finished and before you leave the lab.

QUESTIONS

• CaCO₃ and MgCO₃ are only slightly soluble in water, so they continue to act as micro-grit abrasive when wet. Na₂CO₃ and sodium carbonate peroxide are soluble in water and show no abrasive effect. Sodium carbonate peroxide generates H₂O₂, a cleaner.

Disposal

• Paper cups and pH paper strips can be discarded in the trash. Liquids can be washed down the sink with lots of water.

1. Analyzing Data Using your data from **Data Table 1,** which toothpaste type has a higher scratch index—pastes, gels, or drops?

Answers will vary. Generally, gels should score lower.

2. Applying Conclusions Brushing with ordinary baking soda (sodium bicarbonate, $NaHCO_3$) has been recommended if toothpaste is unavailable. While baking soda is not an abrasive, it does provide a benefit as a dentifrice. What is the benefit?

Its high pH neutralizes mouth acids, thus promoting

remineralization.

1. **Inferring Conclusions** Hydroxyapatite, $Ca_5 (PO_4)_3OH$, is a complex three-dimensional structure in which a series of calcium ions (Ca^{2+}), phosphate ions (PO_4^{3-}), and hydroxide ions (OH^-) are positioned around one-another. See **Figure C.** What is the net charge on one formula unit for this salt?

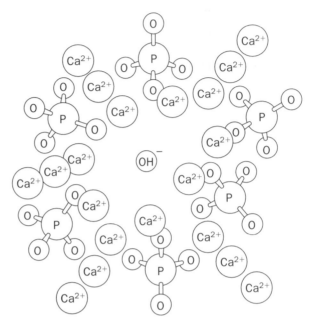

The lattice structure of hydroxyapatite

FIGURE C

The salt has a net charge of 0: there are 5 Ca^{2+} for every 3 PO_4^{3-}

and every 1 OH^- in the lattice.

Name_____

Date_____ Class_____

HOLT
ChemFile
LAB PROGRAM

Matter and Change: Pizza Mixture

NOTE: This experiment should be done at home or in the home economics lab at school.

OBJECTIVES

Recommended time:
2 h

- **Demonstrate** proficiency in identifying and classifying a mixture.
- **Identify** methods for separating a mixture.

INTRODUCTION

Many of the foods we eat are mixtures. Pizza, for example, is a mixture of dough, sauce, cheese, and toppings. In this experiment, you will prepare a pizza and examine it for evidence that it is a mixture.

MATERIALS

Materials

If this lab is performed at school, refer to the box label for items needed that are not contained in the mix.

- 1 box pizza mix, or equivalent
- 1 pizza pan
- 1 pizza cutter or knife
- 1 or 2 hot mitts
- oven

PROCEDURE

Procedural Tips

● This lab is designed to be a home lab assignment. However, if your school offers home economic courses, you may want to arrange to use the kitchen facility.

Pre-lab Discussion

Students should be familiar with the definition of a mixture and with the types of mixtures.

Disposal

No special requirements for disposal are needed.

1. Prepare the pizza crust according to the instructions on the box. Add toppings, and bake according to the instructions.

2. Use a hot mitt to remove the pizza from the oven, and turn the oven off.

3. Cut the pizza into slices. Choose one slice to examine as you answer the post-lab questions.

Cleanup and Disposal

4. Clean your work area and all apparatus. Return equipment to its proper place. Wash hands thoroughly after all work is finished.

QUESTIONS

1. **Analyzing Ideas** Classify the pizza as a heterogeneous mixture or a homogeneous mixture. Explain your reasoning.

   heterogeneous; composition varies throughout the pizza slice.

   _____

   _____

2. **Relating Ideas** Explain how the distribution of toppings indicates that pizza is a mixture.

Toppings are not present in definite ratios. Consider any answer

that indicates an understanding of the definition of a mixture.

GENERAL CONCLUSIONS

Because collodial suspensions are on the dividing line between solutions and heterogeneous mixtures, some students may classify a marshmallow as homogeneous.

1. **Analyzing Methods** Think about how you would remove a topping you do not like. How does this action indicate that pizza is a mixture?

Separation is accomplished by physical means.

2. **Applying Concepts** Powdered lemonade dissolves in water. Would the resulting solution be a mixture or a pure substance? Give a reason for your answer.

mixture; Answers will vary but should indicate that the composi-

tion of the solution is the same throughout, but the water mole-

cules are separate and distinct from lemonade particles.

3. **Applying Concepts** Marshmallows are formed by whipping a sugar mixture into egg whites. The air that is trapped during the whipping action gives the spongy feel to the marshmallow. Is a marshmallow a heterogeneous mixture or a homogeneous mixture? Explain your reasoning and identify the phase or phases present in a marshmallow.

heterogeneous; Answers will vary but should support the type of

mixture chosen. The phases present are a gas dispersed in a solid

to form a solid foam colloid.

EXPERIMENT D5

A Close Look at Aspirin

OBJECTIVES

Recommended time:
45 minutes

- **Test** for the presence of pharmacological fillers in aspirin tablets.
- **Analyze** aspirin tablets for effectiveness.
- **Compare** cost-benefit relationships among tablet brands.

INTRODUCTION

Solution/Material Preparation

1. Wear safety goggles, a face shield, impermeable gloves, and an apron when you prepare the HCl solution. To prepare 1 L of 1.0 M HCl solution, observe the required safety precautions. Slowly and while stirring, add 86 mL of 12 M HCl to 500 mL of distilled water. Dilute to 1 L.

Aspirin is the common name for acetylsalicylic acid. It is easily prepared by reacting acetic anhydride with salicylic acid. Neither of these compounds occurs in nature. Acetic anhydride is made by removing a water molecule from acetic acid, a petroleum distillate. The bark of the willow tree contains salicin, itself an analgesic, which can be converted to salicylic acid.

Most aspirin tablets contain between 300 and 350 mg of acetylsalicylic acid. Extra-strength tablets each contain 500 mg or sometimes even larger amounts. Approximately 650 mg of aspirin has a potency (effectiveness) equal to 32 mg of codeine.

Although acetylsalicylic acid is the active ingredient in aspirin, it is only a portion of what an aspirin tablet contains. Drug companies add fillers, such as starch and glucose, to "bulk up" tablets and to act as binders that aid in holding the tablet's form. Aspirin does not begin to do its work until it enters the bloodstream. How fast this delivery system is depends on the rate at which it disintegrates in the stomach. Certain brands have additional compounds that aid tablet disintegration. Buffered aspirin has a combination of aspirin and one or more bases, such as $MgCO_3$ and aluminum glycinate. It is compounded to disintegrate at a faster rate than regular nonbuffered aspirin.

SAFETY

2. To prepare 100 mL of saturated $NaCO_3$ solution, observe the required safety precautions. Add 30 g of $NaCO_3$ to 100 mL of distilled water. Continue adding $NaCO_3$ until it no longer dissolves.
3. To prepare 1 L of quantitative Benedict's solution, dissolve 200 g of crystalline $NaCO_3$, 200 g of sodium citrate, and 125 g of KSCN in about 800 mL of distilled water. Warm the solution, and filter if necessary. Now dissolve 18 g of crystallized $CuSO_4$ in about 100 mL

  Always wear safety goggles and a lab apron to protect your eyes and clothing. If you get a chemical in your eyes, immediately flush the chemical out at the eyewash station while calling to your teacher. Know the location of the emergency lab shower and the eyewash station and the procedures for using them.

 Do not touch any chemicals. If you get a chemical on your skin or clothing, wash the chemical off at the sink while calling to your teacher. Make sure you carefully read the labels and follow the precautions on all containers of chemicals that you use. If there are no precautions stated on the label, ask your teacher what precautions you should follow. Do not taste any chemicals or items used in the laboratory. Never return leftovers to their original containers; take only small amounts to avoid wasting supplies.

 Call your teacher in the event of a spill. Spills should be cleaned up promptly, according to your teacher's directions.

of distilled water, and pour this solution into the citrate-carbonate-thiocyanate solution slowly while stirring constantly. Dilute this solution to exactly 1 L. 25 mL of this reagent is reduced by 50 mg of glucose.

MATERIALS

Never put broken glass in a regular waste container. Broken glass should be disposed of properly.

When using a hot plate, do not heat glassware that is broken, chipped, or cracked. Use tongs or a hot mitt to handle heated glassware and other equipment because hot glassware does not always look hot.

- 5 different generic and name-brand aspirin tablets
- Benedict's solution
- 1% $FeCl_3$ solution
- Lugol's iodine
- saturated $NaCO_3$
- 10 mL graduated cylinder
- 25 mL graduated cylinders
- 500 mL beaker
- centigram balance
- hot plate

- medicine droppers or micropipets
- microspatula
- mortar and pestle
- plastic (1 oz) cups
- test-tube clamp
- test-tube rack
- test tubes
- universal pH test paper
- wax pencil

PROCEDURE

4. To prepare 100 mL of Lugol's iodine, observe the required safety precautions. Dissolve 10 g of KI in 100 mL of distilled water; then add 5 g of I_2.

5. To prepare a 1% $FeCl_3$ solution, observe the required safety precautions. Dissolve 1 g of $FeCl_3$ in 100 mL of distilled water.

6. Select five different brands/types of aspirin tablets. Label each container with the purchase price. Set the containers in a shared reagent area where students can obtain their tablets and consumer information. Select noncoated tablets only; *do not use caplets.* Tablet selections should include: 81 mg baby aspirin, 325 mg generic and name-brand aspirin, 500 mg generic and

Part 1: Testing for the presence of tablet fillers

1. Obtain a set of five cups and label them 1 to 5. Place three tablets of a different brand/type of aspirin in each labeled cup. Record the manufacturer's name and the strength (amount) of acetylsalicylic acid in **Data Table 1.** The amount of active ingredient is reported in either grains or milligrams. Complete **Calculations** item 1.

2. For each sample tested, determine the mass to the nearest 0.001 g of a single tablet. Record these values in **Data Table 1.** Complete **Calculations** items 2 and 3.

3. Prepare a water bath by adding approximately 250 mL of tap water to a 500 mL beaker. Set the beaker on a hot plate. Heating on medium, bring the water to a slow boil. Complete steps 4 through 10 while waiting for the water to boil.

Part 2: Testing for filler/binder components

4. Using the wax pencil, label five test tubes 1-1, 2-1, 3-1, 4-1, and 5-1. Place the tubes in a test-tube rack.

5. Label ten small cups 1-2, 1-3, 2-2, 2-3, 3-2, 3-3, 4-2, 4-3, 5-2, and 5-3.

6. Measure 15 mL of distilled water in a graduated cylinder. Crush 2 tablets of sample 1 using a mortar and pestle. Use a microspatula to transfer the powdered material to the graduated cylinder. Carefully place your thumb over the top of the cylinder and shake gently to dissolve the aspirin. Pour 5 mL of the mixture into test tube 1-1 and 5 mL each into cups 1-2 and 1-3.

7. Repeat step 6 for brands 2 through 5.

8. pH Test Dip a piece of pH paper into one of the cups from Step 6. Record the value in **Data Table 1.**

name-brand extra-strength aspirin and 325 mg regular and name-brand buffered aspirin. (**Note:** For buffered brands, students should have access to the manufacturer's complete active ingredient(s) label statement, e.g., "a formulation buffered with calcium carbonate, magnesium oxide, and magnesium carbonate.")

Required Precautions

• Read all safety precautions, and discuss them with your students.
• Safety goggles and an apron must be worn at all times.
• In case of an acid or a base spill, first dilute with water. Then mop up the spill with wet cloths or a wet cloth mop while wearing disposable plastic gloves. Designate separate cloths or mops for acid and base spills.
• Broken glass should be disposed of in a clearly labeled box lined with a plastic trash bag. When the box is full, close it, seal it with packaging tape, and set it next to the trash can for disposal.

9. **Sugar Test** To test tube 1-1, add a single drop of saturated $NaCO_3$. Gently swirl the test tube to mix. Use universal pH paper to test for a neutral pH. Repeat the drop-by-drop addition of $NaCO_3$ solution until the pH equals 7.0.

10. Add 5 mL Benedict's quantitative solution to the test tube from step 9.

11. Using a test-tube clamp, carefully place the test tube from step 9 in the water bath for at least 5 minutes.

12. Using a test-tube clamp, remove the test tube from the water bath, and place it in the test-tube rack. If the color of the mixture changes to red, orange, or yellow, reducing sugars are present. Place a check mark next to "glucose" in **Data Table 1.** The darker the color is, the greater the amount of sugar present. A precipitate may also be present for a positive test.

13. Repeat steps 8 through 12 for the other aspirin samples.

14. **Test for Starch** Add five drops of Lugol's iodine to each of the samples in cups 1-2, 2-2, 3-2, 4-2, and 5-2. Gently swirl each cup to mix the iodine solution and the sample. If a blue-black color forms, place a check mark next to "Starch" in **Data Table 1.**

15. **Test for Excess Salicylic Acid** Add 10 drops of 1% $FeCl_3$ to each of the samples in cups 1-3, 2-3, 3-3, 4-3, and 5-3. Gently swirl each cup to mix the $FeCl_3$ reagent and the sample. A pink, light purple, or green color indicates the presence of excess salicylic acid. Record your results in **Data Table 1.**

Data Table 1—*Aspirin Analysis*

Sample number	Strength of active ingredient (g)	Mass of tablet (g)	Percentage filler/binder	pH	Composition of filler/binder	Presence of excess salicylic acid	Time needed to dissolve (min)
Brand 1	0.325	0.531	38.8	< 7	Starch [√] Glucose [ ]	Yes [ ] No [√]	5
Brand 2	0.081	0.140	42.1	< 7	Starch [√] Glucose [√]	Yes [ ] No [√]	1.30
Brand 3	0.081	0.131	38.2	< 7	Starch [√] Glucose [√]	Yes [√] No [ ]	1.28
Brand 4	0.325	0.525	38.1	< 7	Starch [√] Glucose [ ]	Yes [ ] No [√]	5.30
Brand 5	0.500	0.801	37.6	< 7	Starch [√] Glucose [ ]	Yes [ ] No [√]	7

• Caution students to exercise great care when handling heated materials.

Techniques to Demonstrate

• You may want to show students proper use of a mortar and pestle.

Procedural Tips

• Most extra-strength and enteric aspirin tablets are coated. If you use a coated brand, be sure all tablets are coated and that you test several tablets prior to class to be sure the tablets will dissolve in a reasonable amount of time.
• Data in Data Table 1 will vary according to type and age of tablet. Coated tablets take considerably longer than noncoated tablets. Extra-strength tablets are generally coated.
• Data in Data Table 2 is example only and may not be representative of your local area.

Part 3: Analyzing aspirin tablets for effectiveness

16. Use a wax pencil to label five clean, dry test tubes 1 to 5. Place the tubes in a test-tube rack.

17. Pour 10 mL of 1 M HCl into each test tube.

18. Place a single tablet of each sample to be tested into its respective numbered test tube. Record the time it takes to fully dissolve in **Data Table 1.** Why is 1 M HCl, rather than some other acid, used to dissolve the tablets?

1 M HCl most closely models human stomach acid.

Part 4 Determining cost effectiveness

19. Obtain from your teacher consumer information sources for generic, name brand, and extra-strength aspirin tablets. Use this information to complete columns 1, 2, 3, and 5 in **Data Table 2.** Complete **Calculations** steps 4 and 5.

Cleanup and Disposal

20. Clean all apparatus and your lab station. Return equipment to its proper place. Dispose of chemicals and solutions in the containers designated by your teacher. Do not pour any chemicals down the drain or put them in the trash unless your teacher directs you to do so. Wash your hands thoroughly after all work is finished and before you leave the lab.

Data Table 2—*Analyzing and Comparing Tablet Cost vs. Value*					
Brand	**Number of tablets per package**	**Package cost**	**Cost per tablet**	**Amount of active ingredient (mg)**	**Cost per mg of active ingredient**
Bayer	100	$5.95	$0.60	325	0.002 cents
St. Joseph	36	$2.20	$0.06	81	0.0008 cents
Walgreen's	36	$1.33	$0.04	81	0.0005 cents
Walgreen's	300	$3.99	$0.01	325	0.00004 cents
BC Powder	6	$2.39	$0.40	650	0.0006 cents

CALCULATIONS

1. **Organizing Data** Convert the mass of the reported active ingredient for each sample to grams. Enter each result in column 1 of **Data Table 1.** 1 mg = 0.001 g, and 1 gr = 64.8 mg. *Note*: A grain (gr) is the smallest apothecary measure of weight.

   If sample A has 5 gr of acetylsalicylic acid, then

   $$5 \text{ gr} \times \frac{64.8 \text{ mg}}{\text{gr}} \times \frac{0.001 \text{ g}}{\text{mg}} = 0.324 \text{ g}$$

2. **Organizing Data** Calculate the mass of filler/binder in each tablet by subtracting the mass of the active ingredient from the mass of the tablet.

   Mass of tablet (g) – mass of acetylsalicylic acid (g) = mass of filler/binder

3. **Organizing Data** Determine the cost per tablet for each type of aspirin tablet by dividing the package cost by the number of tablets per package. Enter your results in column 4 of **Data Table 2.**

   For a 100 tablet package that costs $5.99, the cost per tablet is $0.06.

   $$\frac{\$5.99}{100 \text{ tablets}} = \$0.0599 \text{ per tablet}$$

4. **Organizing Data** To determine the cost per milligram of active ingredient for each tablet, divide the cost per tablet by the number of milligrams of active ingredient. Enter your results in column 6 of **Data Table 2.**

   If the cost per tablet is $0.06 and the amount of salicylic acid is 325 mg, the cost per milligram of active ingredient is 0.0002 cents.

   $$\frac{\$0.06}{325 \text{ mg}} = 0.0002 \text{ cents/mg}$$

QUESTIONS

1. **Applying Conclusions** Explain why the pH measurements of the tested aspirin tablets are so close together.

   The majority of the compound in each tablet is acetylsalicylic

   acid.

2. **Analyzing Data** Which filler predominates in the aspirin tablets tested, starch or glucose?

   generally starch

3. **Analyzing Data** How uniform are the masses of tablets for aspirin tablets having 325 mg of active ingredient?

Answers will vary. Students should observe differences in masses

of the samples because different fillers are used.

4. **Analyzing Data and Inferring Conclusions** Suggest an explanation for why excess salicylic acid might be present in aspirin tablets.

Aspirin reacts with water to produce salicylic acid and acetic acid.

Salicylic acid is present in aspirin tablets from an old bottle that

has had repeated exposure to humidity.

5. **Analyzing Data** Review the times recorded for each tablet to dissolve. Which brand was most effective in reaching the bloodstream?

The brand with the lowest recorded time.

GENERAL CONCLUSIONS

Pre-Lab Discussion

• Discuss with students how pellets are formed and the purpose of a binder. In addition to starch and glucose, typical fillers/binders include hydroxypropyl methylcellulose (expands in water, a thickening agent), dibasic calcium phosphate (insoluble in water, a bulking agent), and triacetin or glycerol triacetate (a plastisizer).

Disposal

All student materials may be washed down the sink, with copious amounts of water.

1. **Analyzing Information** Review column 6 in **Data Table 2.** This column is the cost per milligram of acetylsalicylic acid. Is it worth paying more money for extra-strength tablets? How does the mass of acetylsalicylic acid in three regular-strength aspirin tablets compare to that in two extra-strength aspirin tablets?

Answers will vary. Three generic 325 mg tablets (975 mg, total)

cost less than two 500 mg extra-strength tablets (1000 mg). The

25 mg difference is negligible pharmacologically. On a cost/mg

basis, one pays a premium for the extra-strength tablets.

2. **Inferring Conclusions** Why does the packaging of some aspirin tablets list active ingredients other than acetylsalicylic acid?

Under FDA regulations, only ingredients that provide a direct

therapeutic effect can be listed as active. Buffered aspirin can list

inorganic compounds, such as $CaCO_3$, MgO, and $MgCO_3$, as ac-

tive ingredients because they permit faster disintegration of the

tablet, thus effecting a faster transport of the drug into the

bloodstream.

Name_____

Date_____ Class_____

A Cloth of Many Colors

OBJECTIVES

Recommended Time:
45 minutes (Part 1)

- **Observe** how direct dyeing and mordant dyeing color cotton fibers.
- **Identify** a class of compounds by molecular structure.
- **Evaluate** the colorfastness and lightfastness of various dyes and dyeing processes.

INTRODUCTION

Plan additional time for students to wash fabric pieces and analyze fade test results (Part 2). Light fade tests should be planned for a minimum of one week; a longer time period yields more definitive results.

Solution/Material Preparation

- Wear safety goggles, a face shield, imperme-able gloves, and an apron when you pre-pare the alum, dye, Na_2CO_3, and NaOH so-lutions. Avoid dusting conditions. Work in a chemical fume hood known to be in operat-ing condition and have another person stand by to call for help in case of an emergency. Be sure you are within a 30 second walk from a safety shower and eye-wash station known to be in good operating condition.
1. To prepare 1 L of 0.5 M $KAl(SO_4)_2 \cdot 12H_2O$ solution, observe the required safety precau-tions. Add 237 g of $KAl(SO_4)_2 \cdot 12H_2O$ to 200 mL distilled water in a volumetric flask. Gently agitate this mix-ture until all solute is dissolved. Add distilled water to exactly 1 L.

A dye is a colored substance used to impart more or less permanent color to other substances. Dyes color most manufactured products, but their most important use is with textile fibers and fabrics. A dye must have an affinity for the substance it colors; it must enter the interior of the fiber and adhere. Starch and various other materials are applied to fabrics to increase their stiffness. These materials (called sizing) clog the pores of fibers and must be removed before the dye can reach the fibers. When a successful commercial dye resists fading when washed repeatedly, the dye is *colorfast*. When the dye also resists fading when exposed to direct light, it is *lightfast*. Not all dyes are suitable for use with every type of fabric. The chemistry of the fiber is critical to the binding of any dye: wool (polypeptide), cotton (cellulose), nylon (polyamide), dacron (polyester). A dye that is suitable for nylon may not dye cotton at all.

Cotton's fiber chemistry is versatile, making it an ideal medium for dye com-pounds and dyeing processes. Direct dyeing is the simplest process: a dye that is in solution attaches itself to the fiber of a fabric by direct chemical interaction. The majority of synthetic direct dyes are *azo* dyes, compounds that have at least one N=N linkage in the molecule. Diazo dyes have two N=N linkages and are ex-cellent cotton fabric dyes. Interestingly, monoazo dyes, compounds with only one N=N linkage, do not bond well to cotton fiber.

NaO_3S—⟨ ⟩—$N=N$—⟨ ⟩—$N(CH_3)_2$
Methyl orange

Alizarin

NaO_3S—⟨ ⟩—$N=N$—⟨HO⟩
Sunset Yellow SO_3Na

Allura Red

H_2N—⟨ ⟩—$N=N$—⟨ ⟩—$N=N$—⟨ ⟩—$NH_2 \cdot 2HCl$
Bismark Brown Y

FIGURE A: Structures of dyes used in this experiment

2. Dyeing solutions: To prepare a 0.2% solution of each dye, observe the required safety precautions. Add 2 g of the dye to 800 mL distilled water. Gently agitate this mixture until all dye is dissolved. Dilute to 1 L.

3. To prepare 0.25 M Na_2CO_3 solution, observe the required safety precautions. Add 31.0 g of $Na_2CO_3 \cdot H_2O$ to 200 mL distilled water in a volumetric flask. Gently agitate this mixture until all solute is dissolved. Add distilled water to exactly 1 L.

Another dye process is mordanting. The fabric is first treated with a heavy metal salt, the mordant, which binds with cotton fibers. When the dye is applied, it links with the mordant: sort of a "chemical handshake" among three individuals! Typical mordants are alum (potassium aluminum sulfate), copper (II) sulfate, or potassium dichromate. **Figure B** shows how direct and mordant dyes bind to cellulose fibers in cotton fabrics.

FIGURE B

SAFETY

4. To prepare 0.5 M NaOH solution, observe the required safety precautions. Add 20.0 g of NaOH to 400 mL distilled water in a volumetric flask. Avoid dusting conditions. Gently agitate this mixture until all the solute is dissolved. Add distilled water to exactly 1 L.

5. Use 100% white cotton fabric as a source for sample material. Cut the fabric into 5 in. × 5 in. pieces, six pieces per group.

Required Precautions

● Read all safety precautions, and discuss them with your students.
● Safety goggles and an apron must be worn at all times.
● In case of a spill, use a dampened cloth or paper towels to mop up the spill. Then rinse the cloth in running water at the sink, wring it out until it is only damp, and put it in the trash.

Always wear safety goggles and a lab apron to protect your eyes and clothing. If you get a chemical in your eyes, immediately flush the chemical out at the eyewash station while calling to your teacher. Know the location of the emergency lab shower and the eyewash station and the procedure for using them.

Do not touch any chemicals. If you get a chemical on your skin or clothing, wash the chemical off at the sink while calling to your teacher. Make sure you carefully read the labels and follow the precautions on all containers of chemicals that you use. If there are no precautions stated on the label, ask your teacher what precautions you should follow. Do not taste any chemicals or items used in the laboratory. Never return leftovers to their original containers; take only small amounts to avoid wasting supplies.

Call your teacher in the event of a spill. Spills should be cleaned up promptly, according to your teacher's directions.

Never put broken glass in a regular waste container. Broken glass should be disposed of properly.

When using a hot plate, do not heat glassware that is broken, chipped, or cracked. Use tongs or a hot mitt to handle heated glassware and other equipment because hot glassware does not always look hot.

MATERIALS

- Broken glass should be disposed of in a clearly labeled box lined with a plastic trash bag. When full, close the box, seal it with packaging tape, and set it next to the trash can for disposal.
- Caution students to exercise great care when handling heated materials.

PROCEDURE

- Dyes are available through science supply houses.
- Assign each student group one dye solution. A dye may be assigned more than once.

Techniques to Demonstrate

- Show students how to cut 1-in square fabric pieces for distribution.

Procedural Tips

- Plan to have a clothesline for students to clothespin their samples for drying. Make a dishpan available in a sink area so that students can wash dyed fabrics.
- Observe fabric samples in as strong a light as possible. You may choose to have students use fine-point forceps to remove threads from each of the fabric samples, prepare wet mounts, and make observations under low power (40×) of a compound microscope.

- 0.5 M aluminum potassium sulfate
- 0.2% methyl orange solution
- 0.2% alizarin solution
- 0.2% Sunset Yellow solution
- 0.2% Allura Red solution
- 0.2% Bismark Brown Y solution
- 0.25 M Na_2CO_3
- 0.5 M NaOH
- 1 L beakers, 3
- magnifying glass or compound microscope, 10×, including slides, cover slips, and fine-point tweezers

- 5 in. × 5 in. cotton fabric pieces
- clothespins
- clothesline
- dishpan
- hot plate
- indelible ink laundry marker
- utility tongs
- white glue
- washing soap

Part 1: Removing fabric sizing

1. Using an indelible ink laundry marker, label each of the six fabric pieces with *one* of the following labels: Direct C, Mordant C, Direct L, Mordant L, Direct W, or Mordant W.

2. After labeling, place the pieces of cloth in a 1 L beaker. Add enough 0.25 M Na_2CO_3 to cover the cloths.

3. Place the beaker on a hot plate and bring the solution to a boil. Continue heating for 2 minutes. Turn off the heat and allow the solution to cool.

4. Using utility tongs, remove the treated cloths from the solution and rinse them thoroughly in running water. Wring dry. What other treatment could be employed to remove fabric sizing?

   machine washing with laundry detergent

Part 2: Direct Dyeing

5. Select the fabric pieces from Step 4 that are labeled with the word *Direct*. Place them in a 1 L beaker.

6. Obtain from your teacher a liquid dye sample. Carefully pour 300 mL of the dye solution into the 1 L beaker. Place the beaker on a hot plate. Bring the solution to a boil and continue heating for 5 minutes. Allow the solution to cool.

7. Using utility tongs, remove the cloths from the dyeing solution. Rinse them under running water. Wring out the excess liquid. Using clothespins, hang the cloths up to dry. *Save the dye solution for use in step 13.*

Part 3: Mordant Dyeing

8. Select the fabric pieces from step 4 that are labeled with the word *Mordant*. Place them in a second 1 L beaker. Pour 300 mL of 0.5 M $KAl(SO_4)_2 \cdot 12H_2O$ into this beaker.

9. Place the beaker on a hot plate and bring the solution to a boil. Continue heating for 5 minutes. Allow the solution to cool.

Pre-Lab Discussion

• You may wish to point out that Allura red (Red 40) and Sunset yellow (Yellow No. 6) are food additives in Kool Aid® drink mixes.
• Remind students that boiling water can scald. Emphasize that heated liquids need time to cool before removing fabric samples. Always use utility tongs to handle fabrics heated to boiling.
• Students test a single dye, cut their test pieces into 1-in. squares, and share with class members.
• A 10× magnifying glass is suggested for step 19. If possible, prepare wet mount standards as viewing samples for the class. Observation of the individual fibers under a compound microscope should show that mordanted fibers appear coated, while direct-stained fibers appear infiltrated with dye.
    Lightfastness depends on the amount of exposure time.

Disposal

Solutions may be flushed down the drain, using copious amounts of water.

10. Using utility tongs, remove the cloth pieces, and rinse them under running water. Then place the cloth pieces into a third 1 L beaker. Pour 200 mL of 0.5 M NaOH into this beaker.

11. Place the beaker on a hot plate and bring the solution to a boil. Continue heating for 5 minutes. Allow the solution to cool.

12. Using utility tongs, remove the three cloth pieces. Rinse each under running water. Wring dry.

13. Place the cloth pieces from step 12 into the beaker of dye solution you saved in Step 7. Place the beaker on a hot plate, and bring the solution to a boil. Continue heating for 5 min. Allow the solution to cool.

14. Using utility tongs, remove each cloth piece from the solution. Rinse it under running water. Wring out the excess liquid. Using clothespins, hang each cloth up to dry.

Part 4: Evaluating dyeing methods

15. Fill a dishpan 2/3 full of tap water. Dissolve a generous amount of soap in the water. Wash each of the two dried fabric pieces labeled with a "W" *at least* 10 times. Rinse thoroughly between washings. Redry.

16. Place each of the two dried fabric pieces labeled with an "L" on a window sill that has a direct southern exposure. Expose these pieces for at least a week.

17. *Do not* wash or place in direct sunlight the two dried fabric pieces labeled with a "C." What is the importance of these fabric pieces to your evaluations?

They serve as controls.

18. Cut the fabric pieces into 1-in. squares so that each lab group can paste a representative sample in **Data Table 1.**

19. Use a magnifying glass, or compound microscope and slides, to view individual fibers with the representative fabric treatments. Summarize your analysis in **Data Table 2.**

Cleanup and Disposal

20. Clean all apparatus and your lab station. Return equipment to its proper place. Dispose of chemicals and solutions in the containers designated by your teacher. Do not pour any chemicals down the drain or put them in the trash unless your teacher directs you to do so. Wash your hands thoroughly after all work is finished and before you leave the lab.

Data Table 1—*Dyeing Fabrics*

Glue fabric swatches in spaces indicated and compare.

DIRECT Colorfast	**DIRECT** Colorfast	**DIRECT** Colorfast	**DIRECT** Colorfast	**DIRECT** Colorfast
Glue fabric piece here	Glue fabric piece here	Glue fabric piece here	Glue fabric piece here	Glue fabric piece here
Methyl orange	Alizarin	Sunset yellow	Allura red	Bismark brown
DIRECT Lightfast	**DIRECT** Lightfast	**DIRECT** Lightfast	**DIRECT** Lightfast	**DIRECT** Lightfast
Glue fabric piece here	Glue fabric piece here	Glue fabric piece here	Glue fabric piece here	Glue fabric piece here
Methyl orange	Alizarin	Sunset yellow	Allura red	Bismark brown
DIRECT *Control*	**DIRECT** *Control*	**DIRECT** *Control*	**DIRECT** *Control*	**DIRECT** *Control*
Glue fabric piece here	Glue fabric piece here	Glue fabric piece here	Glue fabric piece here	Glue fabric piece here
Methyl orange	Alizarin	Sunset yellow	Allura red	Bismark brown
MORDANT Colorfast	**MORDANT** Colorfast	**MORDANT** Colorfast	**MORDANT** Colorfast	**MORDANT** Colorfast
Glue fabric piece here	Glue fabric piece here	Glue fabric piece here	Glue fabric piece here	Glue fabric piece here
Methyl orange	Alizarin	Sunset yellow	Allura red	Bismark brown
MORDANT Lightfast	**MORDANT** Lightfast	**MORDANT** Lightfast	**MORDANT** Lightfast	**MORDANT** Lightfast
Glue fabric piece here	Glue fabric piece here	Glue fabric piece here	Glue fabric piece here	Glue fabric piece here
Methyl orange	Alizarin	Sunset yellow	Allura red	Bismark brown
MORDANT *Control*	**MORDANT** *Control*	**MORDANT** *Control*	**MORDANT** *Control*	**MORDANT** *Control*
Glue fabric piece here	Glue fabric piece here	Glue fabric piece here	Glue fabric piece here	Glue fabric piece here
Methyl orange	Alizarin	Sunset yellow	Allura red	Bismark brown

Data Table 2—*Analysis Report*

Dye name	Dye type: monoazo, diazo, mordant	Coats fibers (yes/no)	Colors fibers (yes/no)	Colorfast rating: excellent, good, fair	Lightfast rating: excellent, good, fair
Methyl orange	monoazo	no	yes	good/fair	good/fair
Alizarin	mordant	yes	no	excellent/ good	excellent/ good
Sunset Yellow	monoazo	no	yes	fair/poor	fair
Allura Red	monoazo	no	yes	fair/poor	fair
Bismark Brown Y	diazo	no	yes	excellent/ good	excellent/ good

QUESTIONS

Monoazo control swatches should exhibit the most intensity of color when compared with colorfast or light-fast experimental groups. Methyl orange, alizarin, and Bismark Brown Y produce the most intense color. Students should observe only small differences between alizarin and Bismark Brown Y for dye intensity and relative fastness when compared against the respective controls.

1. **Analyzing Methods** Which dye process is most effective at producing colorfast and lightfast dyed fabrics? Give reasons for your answer.

   Students should conclude that mordanting is most effective against color fading. Mordanted fibers should retain the dye as well as direct-dyed fibers. Monoazo dyes (applied as a direct dyeing process) are not direct dyes and will not exhibit the same dyeing properties.

2. **Analyzing Data** View dyeing results. Which dye(s) appear to be the weakest? Which appear to be the strongest?

   Answers will vary, but Bismark Brown Y should be the most intense. Allura Red and Sunset Yellow are least intense.

GENERAL CONCLUSIONS

A much greater difference is observed between these two dyestuffs and methyl orange. Sunset Yellow and Allura Red dyes do not bond well to cotton and do not exhibit fastness.

1. **Predicting Outcomes** If you were the research chemist in a company that planned to market home-use dye products, how would you proceed?

   A product design that incorporates a direct dyestuff, capable of dyeing batches that can be heated over a stove, would most likely be commercially successful based on both cost and results. Mordanting involves more chemicals, is time consuming, and yields results that are not demonstrably superior to that of direct dyeing.

EXPERIMENT D7

All Fats Are Not Equal

OBJECTIVES

Recommended time:
45 minutes

INTRODUCTION

Solution/Material Preparation

1. Wear safety goggles, disposable polyethylene gloves, and an apron when you prepare the iodine tincture solution. Work in a chemical fume hood known to be in operating condition and have another person stand by to call for help in case of an emergency. Be sure you are within a 30 second walk from a working safety shower and eyewash station.
2. To prepare 100 mL of tincture of iodine, add 7 g of I_2 crystals and 5 g of KI to 5 mL of distilled water, and dilute to 100 mL with denatured ethyl alcohol.

- **Determine** the degree of unsaturation in fatty acids.
- **Relate** how melting point indicates the degree of saturation.

In a saturated fatty acid, each carbon atom is connected to its neighbors by single bonds, while in an unsaturated fatty acid, some carbon atoms are connected by two bonds. The number of carbon-carbon double bonds in a molecule is the substance's degree of unsaturation. The degree of unsaturation and the total number of carbon atoms in the fatty acid chains determine the differences between fats and oils. For example, myristic acid is a solid at room temperature according to the **Information Table,** but oleic acid, which has one carbon-carbon double bond, is a liquid. Similarly, you should notice an increase in melting points as you move from myristic acid to stearic acid because the number of carbon atoms increases. In general, at room temperature, fats are solids and oils are liquids. Therefore, you might predict a fat to be mostly saturated fatty acids and an oil to be mainly unsaturated fatty acids.

To determine the degree of unsaturation, scientists test for the amount of iodine that reacts with a 100 g sample of fat or oil. This value is the iodine number. The higher the value of the iodine number, the greater the amount of unsaturation in the fat or oil. When I_2 is added to the colorless fat or oil, the mixture appears red violet, like I_2. During the reaction, the color of the mixture fades as I_2 adds to the carbon-carbon double bond, producing a colorless product.

Information Table—*Representative Fatty Acids of Dietary Fats and Oils*

Fatty acid	Melting point (°C)	Class (saturated or unsaturated)	Molecular structure
Myristic acid	58	Saturated	$CH_3-(CH_2)_{12}-CO_2H$
Palmitic acid	63	Saturated	$CH_3-(CH_2)_{14}-CO_2H$
Stearic acid	71	Saturated	$CH_3-(CH_2)_{16}-CO_2H$
Oleic acid	16	Monounsaturated	$CH_3-(CH_2)_7-CH=CH-(CH_2)_7-CO_2H$
Linoleic acid	−5	Polyunsaturated	$CH_3-(CH_2)_4-CH=CH-CH_2-CH=CH-(CH_2)_7-CO_2H$
Linolenic acid	−11	Polyunsaturated	$CH_3-CH_2-CH=CH-CH_2-CH=CH-CH_2-CH=CH-(CH_2)_7-CO_2H$

SAFETY

  Always wear safety goggles and a lab apron to protect your eyes and clothing. If you get a chemical in your eyes, immediately flush the chemical out at the eyewash station while calling to your teacher. Know the location of the emergency lab shower and the eyewash station and the procedures for using them.

 Do not touch any chemicals. If you get a chemical on your skin or clothing, wash the chemical off at the sink while calling to your teacher. Make sure you carefully read the labels and follow the precautions on all containers of chemicals that you use. If there are no precautions stated on the label, ask your teacher what precautions you should follow. Do not taste any chemicals or items used in the laboratory. Never return leftovers to their original containers; take only small amounts to avoid wasting supplies.

 Call your teacher in the event of a spill. Spills should be cleaned up promptly, according to your teacher's directions.

 Never put broken glass in a regular waste container. Broken glass should be disposed of properly.

MATERIALS

• **CAUTION:** Tincture of iodine is a flammable liquid and a powerful stain. Avoid open flames or sparks. Have students handle with care.
• A single graduated cylinder can be used if it is washed and rinsed between samples to avoid mixing of oils.

- 1 tablespoon butter
- milk chocolate, 1 in. × 0.25 in. piece
- 10 mL coconut oil
- 10 mL cod liver oil
- 15 mL corn oil
- 1–3 mL tincture of iodine
- 1 tablespoon soft margarine
- 1 tablespoon stick margarine
- 10 mL peanut oil
- 10 mL sunflower oil
- 1 tablespoon vegetable shortening
- 25 mL beakers, 6

- 500 mL beaker
- beaker tongs
- 25 mL graduated cylinders, 5
- hot plate
- spatula
- ring stand
- tablespoon
- medium test tubes, 10
- test-tube rack
- alcohol thermometer
- thermometer clamp
- wax pencil

PROCEDURE

Procedural Tips

• If a change does not occur in the sunflower or cod liver oil (these oils have the highest iodine number and will shift color first) within 10 minutes at room temperature, have students place all the samples in a water bath and apply low heat. Within a few minutes, both these oils should turn colorless.

Part 1: Determining the degree of unsaturation in commercially available oils

1. Use a wax pencil to label five individual test tubes "Peanut oil," "Sunflower oil," "Corn oil," "Cod liver oil," and "Coconut oil."

2. Using a graduated cylinder, measure 10 mL of peanut oil and pour it into the appropriately labeled test tube. Set the test tube in a test-tube rack. Do the same for each of the other four oils.

3. Add two drops of tincture of iodine to each labeled test tube. *Carefully* swirl each test tube to disperse the iodine into small droplets. Return the test tube to the test-tube rack.

4. Let each mixture of oil and iodine stand for at least 10 minutes. Note the time it takes for any color change to occur *after* adding the iodine. Record both the time and color change in **Data Table 1.**

5. Determine an "unsaturation ranking" for this set of oil samples based on whether a color change occurs (red violet to colorless). If a color change occurs, record the elapsed time. Record your ranking in **Data Table 1.**

Data Table 1—Determining the Degree of Unsaturation in Oils

Oil type	Number of iodine (I₂) drops	Time to change color (min)	Color change (√)	Analysis ranking (most unsaturated to least) 1–5
Peanut oil	2		√	4
Sunflower oil	2		√	2
Corn oil	2		√	3
Cod liver oil	2	Fastest	√	1
Coconut oil	2	No change	√	5

● It is important to add initially the same amount of iodine to each of the oils. If the addition of iodine does not turn all oils immediately red-violet, continue adding iodine drop by drop.
● Depending on the brand used, some students may rank sunflower oil as having the lowest ranking in Data Table 1.
● In Part 2, it is important to start heating at a temperature below 32°C so that an accurate melting point can be determined for all solids.

Pre-Lab Discussion

● Discuss the iodine number and how it indicates unsaturation. The iodine number is the number of grams of iodine that react with 100 grams of fat. Iodine numbers for the oils in Part 1 are: cod liver oil 135–165, Sunflower oil 125–135, corn oil 110–130, peanut oil 90–100, coconut oil 6–10.

Part 2: Determining the melting point of foodstuffs and the degree of fatty acid saturation

6. Use a wax pencil to label six individual beakers "Vegetable shortening," "Butter," "Corn oil," "Margarine," "Soft margarine," and "Chocolate."

7. Measure a *level* tablespoon (5 g) of each soft food sample, and place it in its correspondingly labeled beaker. Use a spatula to help level each soft food sample. Place the piece of chocolate in its beaker.

8. Using **Figure A** as a guide, prepare a water bath. Place one of the beakers prepared in step 7 in the water bath. Heat on the hot plate's low setting, so that the temperature of the water gradually increases from room temperature. Monitor the temperature. Record the temperature at which the food sample liquefies completely in **Data Table 2.** Using beaker tongs, remove the warmed sample from the water bath. Repeat for each food sample. Record the room temperature for corn oil.

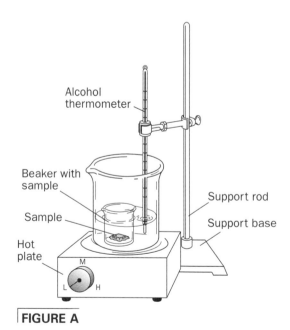

Alcohol thermometer

Beaker with sample

Sample

Hot plate

Support rod

Support base

FIGURE A

9. For each sample tested, review the fatty acid ingredients listed in **Data Table 2** and your melting-point data. Then rank each food sample from highest saturated fatty acid content to lowest saturated fatty acid content. Record this ranking in **Data Table 2.**

Data Table 2—*Melting Point and Degree of Saturation*

Food sample	Melting point (°C)	Fatty acid ingredient(s)	Analysis ranking (most saturated to least)
Vegetable shortening Answers will vary. Margarines should have lower melting points than butter.	22-32	Hydrogenated and partially hydrogenated vegetable oils	5
Butter	About 15	Palmitic acid (29%), oleic acid (27%)	1
Stick margarine	> soft margarine < butter	Partially hydrogenated vegetable oils	2
Soft margarine	< stick margarine	Partially hydrogenated vegetable oils	3
Corn oil	Room temperature	Polyunsaturated acids (34%), oleic acid (50%)	6
Chocolate	About 32	Palmitic acid (24%), stearic acid (35%), oleic acid (38%)	4

● Student's data is qualitative, not quantitative. Although students do not calculate an actual iodine number, the color shift observed and the number of drops of iodine solution used allow an assessment of a generalized ranking. Results may vary according to product brands.

QUESTIONS

● Students may benefit from a discussion of the table in the introduction.
● Review the water-bath setup.

Disposal
Dispose of solid food-stuffs as an inert solid waste. Collect liquid oil mixtures and mix thoroughly with 25 to 50 mL of liquid detergent to disperse the oils, and pour down the drain with copious amounts of water.

Cleanup and Disposal

10. Clean all apparatus and your lab station. Return equipment to its proper place. Dispose of your materials according to your teacher's directions. Dispose of chemicals and waste oils in containers designated by your teacher. Do not pour any chemicals or oils down the drain or put them in the trash unless your teacher directs you to do so. Wash your hands thoroughly after all work is finished and before you leave the lab.

1. **Analyzing Data** Examine your entries in **Data Table 1.** What trend do you observe in vegetable oils regarding unsaturated fatty acid side chains?

   Most vegetable oils are high in unsaturated fatty acids.

2. **Analyzing Data and Applying Concepts** Coconut oil is a major ingredient in many nondairy creamers and other prepared foods. If an individual is trying to reduce saturated fat intake, would a nondairy creamer containing coconut oil be a good choice? Explain.

   No. Cream (butterfat) and coconut oil are both high in

   saturated fatty acids.

EXPERIMENT D8

Polymers as Straws

OBJECTIVES

Recommended time:
45 minutes

- **Measure** the uptake of salt solutions by sodium polyacrylate.
- **Determine** absorbency ratios for the polymer sodium polyacrylate.
- **Graph** the volume of water absorbed versus sodium ion concentration.

INTRODUCTION

Solution/Material Preparation

1. To prepare 1 L of 2.0% NaCl solution, slowly add 20 g of NaCl to 800 mL of distilled water while stirring continuously. Stir until the salt dissolves completely, then add enough distilled water to make 1 L.
2. To prepare 1 L of 1.2% NaCl solution, slowly add 12 g of NaCl to 800 mL of distilled water while stirring continuously. Stir until the salt dissolves completely, then add enough distilled water to make 1 L.

The ability to absorb and retain water plays a critical role in many products from deflocculants to diapers. Recently, polymers have been incorporated into diapers to enhance their absorptive ability. One such "superabsorbent" polymer, sodium polyacrylate, is used in high-absorbency diapers to control leakage. The polymer is marketed under the trade name Waterlock® and is produced by polymerizing sodium acrylate and acrylic acid. Osmotic pressure (the gradient or force that pushes water molecules) causes the superabsorbent polymer to absorb water in an effort to equilize the sodium ion concentration inside and outside the polymer. The concentration of electrolyte in the water being absorbed affects the amount of water the polymer can absorb. The large concentration of sodium inside the polymer makes water flow into the polymer until equilibrium is reached, that is, until the concentration of ions in the polymer equals the concentration of ions outside the polymer. Water is held inside the polymer by hydrogen bonds that form with the sodium acrylate monomers.

The amount of water absorbed by a known amount of polymer is the absorbency ratio. Sodium polyacrylate can absorb approximately 800 times its mass in distilled water (800 g water:1 g sodium polyacrylate). In this experiment, you will add salt water to samples of sodium polyacrylate and determine the absorbency ratio in each situation. By graphing the volume of water absorbed versus the Na$^+$ ion concentration of the salt solutions you will see the effect of salt on the absorbing power of sodium polyacrylate.

SAFETY

3. To prepare 1 L of 0.8% NaCl solution, slowly add 8 g of NaCl to 800 mL of distilled water while stirring continuously. Stir until the salt dissolves completely, then add distilled water to 1 L.

  Always wear safety goggles and a lab apron to protect your eyes and clothing. If you get a chemical in your eyes, immediately flush the chemical out at the eyewash station while calling to your teacher. Know the location of the emergency lab shower and the eyewash station and the procedures for using them.

 Do not touch any chemicals. If you get a chemical on your skin or clothing, wash the chemical off at the sink while calling to your teacher. Make sure you carefully read the labels and follow the precautions on all containers of chemicals that you use. If there are no precautions stated on the label, ask your teacher what precautions you should follow. Do not taste any chemicals or items used in the laboratory. Never return leftovers to their original containers; take only small amounts to avoid wasting supplies.

Call your teacher in the event of a spill. Spills should be cleaned up promptly, according to your teacher's directions.

 Never put broken glass in a regular waste container. Broken glass should be disposed of properly.

MATERIALS

4. To prepare 1 L of 0.4% NaCl solution, slowly add 4 g of NaCl to 800 mL of distilled water while stirring continuously. Stir until the salt dissolves completely, then add enough distilled water to make 1 L.

5. To prepare 1 L of 0.2% NaCl solution, slowly add 2 g of NaCl to 800 mL of distilled water while stirring continuously. Stir until the salt dissolves completely, then add enough distilled water to make 1 L.

- 2.0% NaCl
- 1.2% NaCl
- 0.8% NaCl
- 0.4% NaCl
- 0.2% NaCl
- sodium polyacrylate
- 7 oz clear plastic cups or Petri dishes
- 10 mL graduated cylinder
- 100 mL graduated cylinder
- balance
- micropipet
- microspatula (microspoon)
- wax pencil
- weighing boat

PROCEDURE

6. Set the sodium polyacrylate in a common work area. If sodium polyacrylate is unavailable, it can be obtained by removing it from disposable diapers.

Required Precautions

- Read all safety precautions, and discuss them with your students.
- Safety goggles and a lab apron must be worn at all times.
- In case of a spill, use a dampened cloth or paper towels to mop up the spill. Then rinse the cloth in running water at the sink, wring it out until it is only damp, and put it in the trash.

Techniques to Demonstrate

- Demonstrate how to add the salt solution and determining when to stop adding salt solution.

Part 1: Measuring the uptake of various solutions by sodium polyacrylate

1. Use a wax pencil to label a set of clear plastic cups as follows:

Sample		
1	0% NaCl solution (distilled water)
2	low % NaCl solution (tap water)
3	0.2% NaCl solution
4	0.4% NaCl solution
5	0.8% NaCl solution
6	1.2% NaCl solution
7	2.0% NaCl solution

2. Add 0.1 g of sodium polyacrylate polymer to each cup.

3. Use a 100 mL graduated cylinder to measure 100 mL of distilled water for sample 1.

4. With a micropipet, add sample 1 (distilled water) drop by drop to the polymer sample in cup 1. Add drops until the white powder is no longer visible. *You may need to carefully observe the underside of the cup to make sure no powder is visible.*

5. When no more powder is visible, empty the unused portion from the pipet back into the graduated cylinder. Record the volume of solution remaining in the graduated cylinder in the **Data Table.** Also record your observations of the wetting process.

6. Rinse the graduated cylinder, and repeat steps 3–5 for each of the remaining six sample solutions listed above.

Cleanup and Disposal

7. Clean all apparatus and your lab station. Return equipment to its proper place. Dispose of chemicals and solutions in the containers designated by your teacher. Do not pour any chemicals down the drain or put them in the trash unless your teacher directs you to do so. Wash your hands thoroughly after all work is finished and before you leave the lab.

Data Table 1—*Sodium Polyacrylate Absorbency*					
Sample number	NaCl solution	Volume to wet polymer (mL)	Volume of leftover solution (mL)	Absorbency ratio (mL:g)	Observations
1	0% NaCl solution	70	30	700:1	Wets rapidly
2	Low % NaCl solution	52	48	520:1	
3	0.2% NaCl solution	42	58	420:1	
4	0.4% NaCl solution	30	70	300:1	
5	0.8% NaCl solution	12	88	12:1	
6	1.2% NaCl solution	7	93	7:1	
7	2.0% NaCl solution	2	98	2:1	Wets slowly

Pre-Lab Discussion

● Point out to students that a more accurate data set can be obtained if multiple measurements are taken for each solution concentration and then averaged.
● Emphasize to students that the polymer is thoroughly wetted just as the white powder polymer disappears.

Disposal

Pour salt solutions down the drain. Put the gelled polymer in the trash.

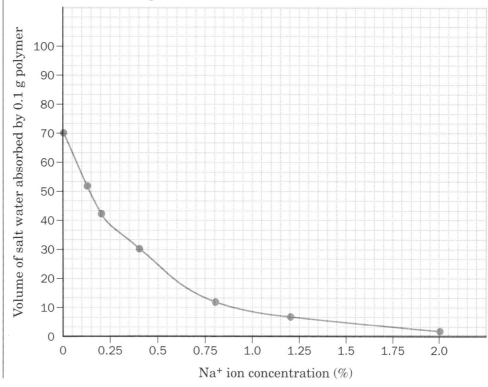

Absorbency vs. Na⁺ Ion Concentration

CALCULATIONS

1. **Organizing Data** Calculate the volume of salt solution absorbed by the polymer. (Hint: Subtract the amount of leftover solution from 100 mL.)

100 mL – 30 mL = 70 mL

2. **Organizing Data** If the salt solution in each trial is absorbed by 0.1 g of polymer, use a ratio-proportion setup to determine the amount of solution absorbed by 1.0 g of polymer. Record your ratios in the **Data Table.**

For sample 1: $\dfrac{70 \text{ mL salt solution}}{0.1 \text{ g polymer}} = \dfrac{x \text{ mL salt solution}}{1.0 \text{ g polymer}}$

$x = \dfrac{(70)(1.0)}{(0.1)} \text{ mL} = 700 \text{ mL}$

QUESTIONS

1. **Interpreting Graphics and Applying Concepts** Urine's salt concentration is approximately 0.9%. Calculate the *minimum* amount of sodium polyacrylate that a diaper must contain to absorb 100 mL.

$\dfrac{0.1 \text{ g polymer}}{10 \text{ mL 0.9\% NaCl solution}} = \dfrac{x \text{ g polymer}}{100 \text{ mL 0.9\% NaCl solution}}$

$x = 1.0$ g polymer

2. **Interpreting Graphics and Applying Concepts** If a disposable diaper contains 1.5 g of sodium polyacrylate, how much urine (0.9% NaCl solution) can it hold?

$\dfrac{0.1 \text{ g polymer}}{10 \text{ mL NaCl solution}} = \dfrac{1.5 \text{ g polymer}}{x \text{ mL NaCl solution}}$

$x = 150$ mL NaCl solution

3. **Evaluating Data** What is the relationship between polymer absorbency and the water's ion concentration?

Absorbency decreases as the water's ion concentration increases.

4. **Analyzing Method** The published absorption ratio of water to sodium polyacrylate is 800:1. Suggest reasons why your experimental results indicate a lower ratio.

Student data will vary (usually between 600:1 and 700:1). Two

possible reasons are the initial weighing of the dry polymer and

human error in determining the point at which the total dissolu-

tion of polymer (disappearance of the white powder) occurs.

GENERAL CONCLUSIONS

1. **Applying Concepts** Explain how the polymer might be used to determine the hardness or softness of tap water.

Hard water contains more Na, Ca, and Mg ions than does soft

water. A correlation could be made between absorbency and

(hard) water volume. The higher the ion concentration is (hard-

ness), the lower the absorbency.

2. **Analyzing Methods and Designing Experiments** Describe other ways this experiment might be conducted to obtain the same relational information.

Answers will vary. Students might suggest that the amount of

water be kept constant and the amount of polymer necessary to

absorb it be determined.

3. **Inferring Conclusions** Consider the absorption process, then identify the salt solution and the polymer as hypertonic or hypotonic. Explain your reasoning.

The salt solution moves into the polymer until equilibrium is es-

tablished, so the polymer is hypertonic relative to the outside en-

vironment (salt solution), which is hypotonic.

4. **Inferring Conclusions** Would you expect the amount of polymer included in diapers to vary? How might you prove or disprove your hypothesis?

probably; more polymer would be included in diapers for larger

toddlers (exceeding 30 pounds) than for infants. Answers will

vary; one approach would be to cut open diapers of varying sizes

and brands, collect the polymer from each, and determine their

masses.

The Slime Challenge

OBJECTIVES

Recommended time:
45 min

- **Prepare** slime by using sodium borate as a cross-linking agent with various natural and synthetic polymers.
- **Compare** the physical properties of prepared slime variants.
- **Evaluate** which polymer makes the best slime.

INTRODUCTION

Solution/Material Preparation

1. To prepare 1 L of 5% polyvinyl alcohol (PVA) solution, heat approximately 700 mL of distilled water to 60–70°C. To avoid dusting, pre-wet the PVA (50 g) in about 100 mL of cold water, mix to a slurry using a spatula. Slowly add *hot* water, using vigorous magnetic stirring. Once dissolution occurs, add distilled water to a final volume of 1 L. Cover the beaker with clear plastic wrap, and continue heating at 60–70°C for 4 to 6 hours, or until the solution clears. Do not exceed 80°C. The time for dissolution will vary depending on the molecular mass of the material; higher molecular masses have higher viscosities. See "Procedure Note" for additional information.

2. To prepare 1 L of cornstarch solution, slowly sprinkle 50 g of cornstarch powder into 750 mL of *hot* water using vigorous magnetic stirring. Add the powder slowly to avoid dusting and clumping. Once dissolution occurs, add the remaining distilled water to a final volume of 1 L.

Slime® is a trade name for a toy material first marketed commercially in the 1970s. Its success dramatizes just how entertaining polymer chemistry can be. Polymers are chains of macromolecules formed by the union of five or more identical combining units (monomers). A cross-linking agent can be added to enhance a polymer's characteristics. In 1839, Charles Goodyear noticed that if natural rubber and sulfur were heated together they became "linked," allowing the rubber to remain firm and elastic at high temperatures instead of soft and sticky. It is the same for Slime.

In this laboratory activity, you will evaluate how the crosslinking reaction of sodium borate affects certain physical characteristics of the four polymers described in the Information Table. Sodium borate decahydrate, $Na_2B_4O_7 \cdot 10H_2O$, is the salt of a strong base (NaOH) and a weak acid $B(OH_3)$. In water, the weak acid hydrolyzes to form a borate ion, which reacts with a hydroxyl group. Polyvinyl alcohol has hydroxyl groups attached to the main polymer chain, so it can link to the borate ion. Following linkage, most of the lattice structure is filled with water molecules, and the carbon-to-oxygen-to-boron bonds are easily broken and re-formed. **Figure A** shows the crosslinking reaction between polyvinyl alcohol and sodium borate.

FIGURE A

Information Table

Polymer name	Natural or Synthetic	Repeating monomeric unit	Polymer type
Cornstarch	Natural		Polysaccharide
		Cornstarch $$\left[\begin{array}{c} \text{CH}_2\text{OH} \\ \end{array} \right]_n$$ Amylose (27%) Amylopectin (73%)	
Guar gum	Natural		Polysaccharide
		Guar gum Galactose (35%) Mannose (65%)	
Polyvinyl alcohol (PVA)	Synthetic		Thermoplastic elastomer
		Polyvinyl alcohol	
Polyvinyl acetate PVAc	Synthetic		Thermoplastic elastomer
		Polyvinyl acetate	

Polysaccharide—a combination of nine or more monosaccharides (simple sugars, $C_6H_{12}O_6$) linked together.

Thermoplastic elastomer—a high-chain polymer that softens when exposed to heat and has the ability to be stretched to twice its original length and to retract rapidly when released.

ChemFile

SAFETY

3. To prepare 1 L of white-glue solution, mix 500 mL of white glue with 300 mL of distilled water, using vigorous magnetic stirring. Add additional distilled water to bring the final volume to 1 L.
4. To prepare 1 L of guar gum solution, slowly sprinkle 17 g of guar gum powder into 800 mL of *warm* water using vigorous magnetic stirring. Add the powder slowly to avoid dusting and clumping. Once dissolution occurs, add distilled water to a final volume of 1 L.
5. To prepare 500 mL of 4% sodium borate solution, add 8 g of sodium borate decahydrate to 300 mL of distilled water, and stir until total dissolution occurs. Add additional distilled water to dilute the solution to 500 mL.
6. For the food coloring, use commercially available colors; green is the most popular slime color.

  Always wear safety goggles and a lab apron to protect your eyes and clothing. If you get a chemical in your eyes, immediately flush the chemical out at the eyewash station while calling to your teacher. Know the location of the emergency lab shower and the eyewash station and the procedures for using them.

 Do not touch any chemicals. If you get a chemical on your skin or clothing, wash the chemical off at the sink while calling to your teacher. Make sure you carefully read the labels and follow the precautions on all containers of chemicals that you use. If there are no precautions stated on the label, ask your teacher what precautions you should follow. Do not taste any chemicals or items used in the laboratory. Never return leftovers to their original containers; take only small amounts to avoid wasting supplies.

 Call your teacher in the event of a spill. Spills should be cleaned up promptly, according to your teacher's directions.

 Never put broken glass in a regular waste container. Broken glass should be disposed of properly.
Never stir with a thermometer because the glass around the bulb is fragile and might break.

 When using a Bunsen burner, confine long hair and loose clothing. If your clothing catches on fire, WALK to the emergency lab shower and use it to put out the fire. Do not heat glassware that is broken, chipped, or cracked. Use tongs or a hot mitt to handle heated glassware and other equipment because hot glassware does not always look hot.

MATERIALS

Required Precautions

- Read all safety precautions, and discuss them with your students.
- Safety goggles and an apron must be worn at all times.
- Although these polymers are made from nontoxic materials, caution students NOT to ingest any of these materials and to wash their hands following handling.

- 4% sodium tetraborate (sodium borate) solution, 10 mL
- 5% polyvinyl alcohol (PVA) solution, 30 mL
- 50% white glue solution, 30 mL
- 1.7% guar gum solution, 30 mL
- cornstarch solution, 30 mL
- 50 mL graduated cylinders, 4
- aluminum pie pans, 2
- 6 in × 2.5 in. piece of card stock
- duct tape
- food coloring, assorted colors

- ice cubes
- iron ring, 3 in.
- 4 oz plastic cups, 8
- ring stand
- 15 cm plastic ruler
- No. 5 solid stopper
- spatula
- stem funnel, 1 in.
- thermometer, nonmercury
- wax pencil
- 8-in. wooden dowel 1/4 in. diameter

PROCEDURE

● Polyvinyl alcohol comes in a variety of molecular masses. The "medium viscosity" (120 000–150 000 g/mol) is best for formulating 3–5% polymer solutions because it does not have to be heated. Higher viscosity (molecular mass) product requires heating. In all cases, pre-wetting the material using cold water to form a slurry is recommended to avoid lumps.
● White glue contains polyvinyl acetate (PVAc).

Procedural Tips

● An additional area of investigation may be to add iron oxide or iron filings—the amounts varying with experimenter's preference—to the slime samples to create Gak,™ another marketed toy polymer. Students can experiment with how Gak responds to various magnetic fields.
● In step 3, although all polymer solutions will flow freely through the funnel, it is important for students to understand that many polymers do not exhibit high viscosity unless they are cross-linked.
● Viscosity of all polymer samples increases greatly when cross-linked. Cross-linked PVA > (PVAc) > guar gum and cornstarch. Cornstarch and guar gum are unaffected by temperature, but flow rates for PVA and PVAc increase when the substances are warm and decrease when the substances are cold.

Pre-Lab Discussion

● Discuss polymerization reactions and what is meant by cross-linking.
● Students should understand the concepts of viscosity and elasticity.

Part 1: Making slime and its variants

1. Use a wax pencil to label two sets of four plastic cups 1 to 4, for a total of eight cups. Then set up a viscometer as illustrated in **Figure B**.

2. Using separate graduated cylinders, pour 30 mL of cornstarch solution into cup 1, 30 mL of guar gum solution into cup 2, 30 mL of polyvinyl alcohol into cup 3, and 30 mL of white glue into cup 4.

3. Carefully dump all the contents of cup 1 into the funnel. Start timing when the liquid first emerges from the funnel's stem. Stop timing when the liquid first reaches the 10 cm mark on your ruler. Record the time in **the Data Table.** Catch the liquid polymer using the second identically numbered cup. Rinse and dry the funnel. Repeat this process for polymer samples 2 through 4.

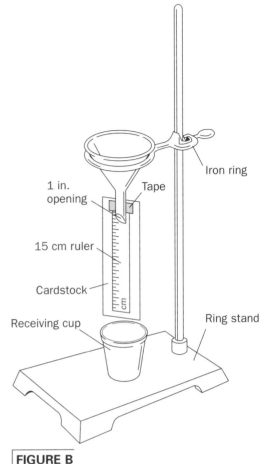

Iron ring
1 in. opening
Tape
15 cm ruler
Cardstock
Receiving cup
Ring stand

FIGURE B

4. Add 1 to 3 drops of food coloring to each polymer solution. To the cornstarch/dye solution in cup 1, add 8 drops of sodium borate solution, and stir the mixture rapidly for 2 minutes. Repeat with the polymer/dye solutions in cups 2, 3, and 4. **Note:** Be sure you add the same number of drops of sodium borate solution in all cases.

Part 2: Evaluating the physical properties of slime variants

5. **Comparing flow rate (viscosity):** Insert a rubber stopper at the bottom of the funnel. Use a wooden dowel to help fill the funnel with slime from cup 1. Carefully remove the rubber stopper and record the time needed for the slime sample to flow out of the funnel stem and past the 10 cm mark. Record the time in **The Data Table.** Recover the slime sample in its numbered cup. Clean and dry the funnel. Repeat for the slime samples in cups 2, 3, and 4.

6. **Comparing elastic recoil:** Carefully insert a wooden dowel into the Slime sample in cup 1. While holding the cup in one hand, move the dowel in a counterclockwise direction with the other hand. Quickly release the dowel. Does the dowel move backwards in a clockwise direction—display elastic recoil? Record your observations in the **Data Table.** Repeat this process for the slime samples in cups 2, 3, and 4.

In step 7, you may wish to have students press their thumbs into each sample and observe what happens. PVA and PVAc deform (stretch) then return to their original shape; cornstarch and guar gum do not deform, so students' fingers go through the material.

7. Take a slime sample in both hands. Slowly stretch it to twice its original length, then let go. Does the sample return rapidly to its original length? Record your results in the **Data Table.** Repeat for the slime samples in cups 2, 3, and 4.

8. Comparing thermal elasticity: Place the cups containing each of the four slime samples in an aluminum pan containing *warm* water. Let the samples set in the water for 5 minutes. Insert a thermometer into the slime sample, and record its temperature in the **Data Table.** Repeat for each warmed slime sample.

9. Place the cups containing each of the four slime samples in an aluminum pan containing *ice* water. Let the samples set in the water for 5 minutes. Insert a thermometer into the slime sample, and record its temperature in the **Data Table.** Repeat for each chilled slime sample.

Cleanup and Disposal

10. Clean all apparatus and your lab station. Return equipment to its proper place. Dispose of chemicals and solutions in the containers designated by your teacher. Do not pour any chemicals down the drain or put them in the trash unless your teacher directs you to do so. Wash your hands thoroughly after all work is finished and before you leave the lab.

Data Table

Sample number	Polymer flow rate (cm/s)	Time (s)	Flow rate (cm/s)	Elastic recoil (yes or no)	Stretch (yes or no)	Thermal elasticity flow rate (cm/s)	
						Heated ____ °C	Cooled ____ °C
1 Cornstarch	Free-flowing	10	1.0	no	no	Time 10-12 s Rate 0.8-1	Time 10-12 s Rate 0.8-1
2 guar gum	Free-flowing	15	0.66	no	no	Time 15 s Rate 0.66	Time 15 s Rate 0.66
3 Polyvinyl alcohol	Free-flowing	30	0.33	yes	yes	Time 18-19 s Rate 0.55	Time 40-41 s Rate 25
4 White glue (PVAc)	Free-flowing	22	0.45	yes	yes	Time 13-14 s Rate 0.77	Time 29-30 s Rate 0.34

CALCULATIONS

1. Organizing Data To calculate flow rate, divide the distance the polymer flowed (10 cm) by the time it took for the polymer to flow this distance. Record your answers in the **Data Table.** Flow rates of less than 1 second are recorded as "free-flowing."

All polymers are free-flowing. Check that calculations are set up correctly.

2. Organizing Data To calculate flow rates for slime samples, divide the distance the polymer flowed (10 cm) by the time it took for the polymer to flow this distance. Record your answers in the **Data Table.**

Answers will vary.

1. Analyzing Data Which polymer material exhibited the highest viscosity? The longest flow rate? Which slime sample exhibited the highest viscosity?

At these concentrations, all are free-flowing liquids. Cross-linked

PVA should exhibit the highest viscosity.

2. Analyzing Data Which polymer sample(s) exhibited the property of elastic recoil?

PVA and PVAc should exhibit the property, PVA being more pro-

nounced than PVAc. The natural polymers do not; they are not

elastomers.

3. Analyzing Data Do all cross-linked polymers rapidly return to their origi-nal length following stretching? Justify your answer using the polymer char-acteristics in the **Information Table.**

No, elastomers PVA and PVAc exhibit stretch elasticity;

nonelastomers do not.

4. Analyzing Data Summarize flow-rate data regarding heating and cooling of slime samples.

Among synthetic polymers, warmer temperatures increased flow

rates and cooler temperatures decreased flow rates. Natural poly-

mers are not thermoplastic, so their flow rates should remain rela-

tively unchanged (temperature independent).

ChemFile

Counting Calories

OBJECTIVES

Recommended time:
45 minutes

- **Construct** a calorimeter setup.
- **Determine** the amount of energy, in Calories, of oil-roasted walnuts and peanuts.
- **Compare** measured caloric data.
- **Determine** which food is a better source of energy.

INTRODUCTION

We expend energy in three ways—through exercise, Specific Dynamic Action (SDA), and through basal metabolism (BM). Exercise is the composite of all physical work we do with our bodies. SDA accounts for the energy consumed in digesting and metabolizing food. BM accounts for the energy needed to maintain key bodily functions that sustain life.

Food provides energy. Food chemicals are classified as fats and oils, carbohydrates, or proteins. Fats furnish 9 Cal per gram, carbohydrates 4 Cal per gram, and proteins 4 Cal per gram. The amount of energy each of these chemicals delivers to the body is independent of the kind of food or quantity of other macronutrients. A meal containing 10 g of fat, 15 g of carbohydrates, and 20 g of protein would deliver 230 Cal of energy: 90 Cal from fat, 60 Cal from carbohydrates, and 80 Cal from protein. In this experiment, you will construct a calorimeter and indirectly determine the caloric content of peanuts and walnuts by measuring the temperature change in a sample of water.

Note: The C in nutritional Calorie is capitalized. The nutritional Calorie is equal to 1000 calories, or 1 kilocalorie.

SAFETY

Required Precautions

- Read all safety precautions, and discuss them with your students.
- Safety goggles and an apron must be worn at all times.
- A beaker filled with water should be placed at each bench setup to extinguish any uncontrolled burning.

Always wear safety goggles and a lab apron to protect your eyes and clothing. If you get a chemical in your eyes, immediately flush the chemical out at the eyewash station while calling to your teacher. Know the location of the emergency lab shower and the eyewash station and the procedures for using them.

Do not touch any chemicals. If you get a chemical on your skin or clothing, wash the chemical off at the sink while calling to your teacher. Make sure you carefully read the labels and follow the precautions on all containers of chemicals that you use. If there are no precautions stated on the label, ask your teacher what precautions you should follow. Do not taste any chemicals or items used in the laboratory. Never return leftovers to their original containers; take only small amounts to avoid wasting supplies.

Call your teacher in the event of a spill. Spills should be cleaned up promptly, according to your teacher's directions.

● Designate one student from each group to monitor the group's activities and call out unsafe issues or react to extinguish any uncontrolled burning. This individual also times cooling periods to prevent potential scalding.
● Caution students to handle matches with care. The box cover should be closed between strikes.

 Never put broken glass in a regular waste container. Broken glass should be disposed of properly.

Never stir with a thermometer because the glass around the bulb is fragile and might break.

 When using a flame, confine long hair and loose clothing. If your clothing catches on fire, WALK to the emergency lab shower and use it to put out the fire. Do not heat glassware that is broken, chipped, or cracked. Use a hot mitt or pot holder to handle heated glassware and other equipment because hot glassware does not always look hot.

MATERIALS

● The nut samples should be of similar mass.
● A home temperature digital recording device is suitable. Lab digital recording devices, and CBLs and probes are also acceptable.

- peanut halves
- walnut halves
- heavy aluminum foil, 5 × 5 cm
- 250 mL flask
- 100 mL graduated cylinder
- aluminum pie plate
- can opener
- centigram balance
- cork stopper
- digital temperature recorder or thermometer
- knife or single-edge razor blade
- matches
- metal file
- metric ruler
- paper clip
- pot holder
- 16 oz tin can
- tin snips

PROCEDURE

Procedural Tips

• Have students practice assembling and disassembling the calorimeter to prevent accidents and to keep to a minimum the time lost following ignition and burning of the food source.
• Student calculated averages in the Data Table will be 20% to 40% of the published caloric value: 600 Cal/100 g for oil-roasted peanuts and 702 Cal/100 g for oil-roasted walnuts.

Pre-Lab Discussion

• Review the methods of heat transfer, combustion reactions, and accuracy in measurement.

Disposal

• Retorts, pie pans, and holder assemblies can be stored for reuse. Charred remains of food can be thrown out with trash.

1. Use **Figure A** to prepare the retort. Remove the label from the can. Use the can opener to punch four triangular openings in the unopened end (top) of the can and three openings in the side of the can near the top. Use tin snips to cut a viewing hole in the side of the can, beginning at the can's open end: 5 cm high, 3 cm wide at the top, and 4 cm wide at the bottom. Use a file to remove all burrs and sharp edges.

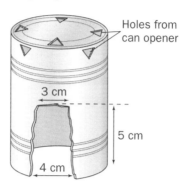

Holes from can opener

3 cm

5 cm

4 cm

FIGURE A

2. Refer to **Figure B** to construct the sample holder. Bend a paper clip so that it can hold food samples between the wires. Measure and cut the cork stopper so that its height when resting on its widest flat surface is 2 cm. Mold a piece of aluminum foil around the cork. Insert the bent paper clip into the cork. The total height of the holder assembly should be no higher than 3.5 to 4.0 cm.

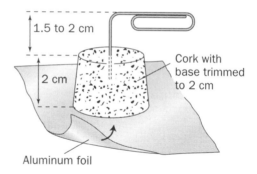

1.5 to 2 cm

Cork with base trimmed to 2 cm

2 cm

Aluminum foil

FIGURE B

3. Obtain three peanut samples and three walnut samples. Use a balance to determine the mass of each sample to the nearest 0.1 g. For each sample, record the type of nut and its mass in the **Data Table.**

4. Using a graduated cylinder, pour 100 mL of tap water into the 250 mL flask.

5. Insert the temperature recorder into the flask. Make sure that the sensing element is fully submerged and does not make contact with the bottom glass. If a digital temperature recording device is unavailable, use an alcohol thermometer.

6. Measure the temperature of the water in the flask; record this value in the **Data Table.**

7. Place a peanut half in the wire holder that is anchored in the cork. Position the wire-holder assembly in the center of the aluminum pie plate.

8. Using matches, carefully set fire to the nut. Once burning is sustained, carefully position the metal can with holes (retort) over the burning sample so that the viewing hole faces you. Set the flask on top of the retort. Refer to **Figure C.**

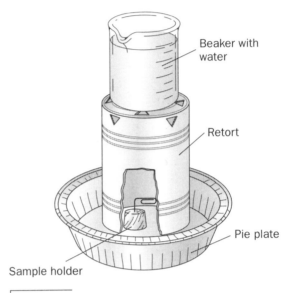

Beaker with water

Retort

Pie plate

Sample holder

| FIGURE C

9. Carefully observe the digital readout of the temperature until it starts to fall. Record the maximum temperature value in the **Data Table.**

10. Cool for 2 to 5 minutes. Using a pot holder, remove the flask from the retort and pour the water into the sink. Using a pot holder, remove the retort from the pie plate.

11. Repeat steps 4 through 9 until you have collected data for three peanut halves and three walnut halves.

Cleanup and Disposal

12. Clean all apparatus and your lab station. Return equipment to its proper place. Dispose of chemicals and solutions in the containers designated by your teacher. Do not pour any chemicals down the drain or put them in the trash unless your teacher directs you to do so. Wash your hands thoroughly after all work is finished and before you leave the lab.

CALCULATIONS

1. **Organizing Data** Determine the average mass for each nut type tested; record this value in the **Data Table.**

Answers will vary. Check that calculations are done correctly. Example data:

Walnuts: $\dfrac{(1.5 \text{ g} + 2.0 \text{ g} + 1.4 \text{ g})}{3} = 1.6 \text{ g}$

Peanuts: $\dfrac{(0.60 \text{ g} + 0.50 \text{ g} + 0.60 \text{ g})}{3} = 0.57 \text{ g}$

ChemFile

Data Table

| | Mass of sample (g) | Water temperature (°C) | | | calories | Food energy | |
		Before burning (T_1)	After burning (T_2)	Change in water temperature		kcal	kcal per 100 g of food sample
Trial 1 peanut	0.60	21.9	30.0	8.1	810	0.810	
Trial 2 peanut	0.50	29.9	35.2	5.3	530	0.530	
Trial 3 peanut	0.60	22.0	30.0	8.0	800	0.800	
Average	0.57			7.1	710	0.710	120
Trial 1 walnut	1.5	26.8	52.7	25.7	2570	2.570	
Trial 2 walnut	2.0	22.0	57.0	35.0	3500	3.500	
Trial 3 walnut	1.4	21.0	46.8	25.8	2580	2.580	
Average	1.6			28.8	2880	2.880	180

2. **Organizing Data** Calculate the change in water temperature for each trial in the **Data Table** by subtracting T_1 from T_2. Enter the results in the **Data Table.** Then calculate the average change in water temperature for each nut type tested; record the values in the **Data Table.**

Example data for peanuts:
For Trial 3, $T_2 = 30.0°C$ and $T_1 = 22.0°C$, so $T_2 - T_1 = 8.0°C$

Peanuts: $\dfrac{(8.1°C + 5.3°C + 8.0°C)}{3} = 7.1°C$

3. **Organizing Data** Determine the Calories (kcal) produced per 100 g of food sample for both nut types tested; record the values in the **Data Table.** The energy transferred by the combustion of the nut equals the energy absorbed by the water. Use 1 cal/g•°C as the specific heat capacity of water when calculating the calories produced.

Example data for peanuts:
$\Delta T = 7.1°C$ (from table)
$7.1°C \times 100$ g $H_2O = 710$ cal

$$710 \text{ cal} \times \frac{1 \text{ kcal}}{1000 \text{ cal}} = 0.710 \text{ kcal}$$

$$\frac{0.710 \text{ kcal}}{0.57 \text{ g (nut's mass)}} = \frac{1.2 \text{ kcal}}{g}$$

$$\frac{1.2 \text{ kcal}}{g} \times \frac{100}{100} = \frac{120 \text{ kcal}}{100 \text{ g}}$$

4. **Predicting Outcomes** Suppose a package of lunchmeat contains 5% fat according to the label. If this value is a percentage of the total mass of the macronutrients, what percentage of the total calories does this amount of fat provide? Assume that the food portion is 100 g and is 5% fat by mass.

The lunchmeat sample contains 5 g of fat and 95 g of carbohydrates and proteins combined. Both carbohydrates and proteins have a caloric value of 4 Cal/g.

Calories from fat: 5 g × 9 Cal/g = 45 Cal

Calories from carbohydrates and proteins:
   95 g × 4 Cal/g = 380 Cal

Total Calories = 45 Cal + 380 Cal = 425 Cal

Percentage of total calories provided by fat:

$$\frac{45 \text{ Cal}}{425 \text{ Cal}} \times 100 = 10.6\%$$

GENERAL CONCLUSIONS

1. **Inferring Conclusions** According to your experimental data, which nut tested is a better source of energy? Why?

Walnuts are a better source of energy because their caloric value

is higher.

Name_____

Date_____ Class_____

Factors Affecting CO_2 Production in Yeast

OBJECTIVES

Recommended time:
2 hours if all three parts of the investigation are done concurrently

- **Distinguish** between aerobic and anaerobic respiration.
- **Calculate** the volume of carbon dioxide gas produced under various conditions.
- **Graph** carbon dioxide production data.
- **Determine** ideal growth conditions for yeast and their relation to carbon dioxide production.

INTRODUCTION

Solution/Material Preparation

1. If an incubator is not available, provide students with an insulating material such as bubble wrap or newspaper to wrap the selected flasks to prevent heat loss.

Fermentation is a chemical process in which microorganisms such as bacteria, molds, and yeasts break down energy-rich organic materials, producing energy, alcohol or water, and carbon dioxide gas. Around 1815, Louis Gay-Lussac, a French chemist, concluded that the production of alcohol and carbon dioxide during fermentation was the result of a physical change in which inert matter decomposed into carbon dioxide and alcohol. This view was accepted until Louis Pasteur proved that fermentation was in fact caused by living microorganisms. Today, fermenters, or bioreactors, are used to manufacture alcoholic beverages, cheese, bread, and pharmaceutical products. In a bioreactor, living cells are mixed with nutrients and grown in a carefully controlled, sterile environment. To make a specific product, the reaction environment's temperature, pressure, pH, oxygen content, and nutrient content must be maintained at levels optimum for the desired product.

In this experiment, you will observe the effect on the natural fermentation process of varying the sugar concentration, temperature, and pH by monitoring and calculating the volume of CO_2 gas produced.

SAFETY

Required Precautions

- Read all safety precautions, and discuss them with your students.
- Safety goggles and a lab apron must be worn at all times.

  Always wear safety goggles and a lab apron to protect your eyes and clothing. If you get a chemical in your eyes, immediately flush the chemical out at the eyewash station while calling to your teacher. Know the location of the emergency lab shower and the eyewash station and the procedures for using them.

 Do not touch any chemicals. If you get a chemical on your skin or clothing, wash the chemical off at the sink while calling to your teacher. Make sure you carefully read the labels and follow the precautions on all containers of chemicals that you use. If there are no precautions stated on the label, ask your teacher what precautions you should follow. Do not taste any chemicals or items used in the laboratory. Never return leftovers to their original containers; take only small amounts to avoid wasting supplies.

• In case of a spill, use a dampened cloth or paper towels to mop up the spill. Then rinse the cloth in running water at the sink, wring it out until it is only damp, and put it in the trash.

Call your teacher in the event of a spill. Spills should be cleaned up promptly, according to your teacher's directions.

Never put broken glass in a regular waste container. Broken glass should be disposed of properly.

Never stir with a thermometer because the glass around the bulb is fragile and might break.

MATERIALS

- sodium carbonate, Na_2CO_3, 15 g
- *saccharomyces cerevisiae* (baker's yeast)
- sucrose (table sugar)
- white vinegar
- 250 mL beakers, 3
- alcohol thermometer
- balance

- small spherical balloons, 9
- Erlenmeyer flasks or glass bottles, 9
- ice-water bath
- incubator or insulating material
- strips of pH paper
- rubber bands, 9
- tape measure
- wax pencil

PROCEDURE

Techniques to Demonstrate

• You may want to show students how to cover the mouth of the flasks and how to secure the balloon with a rubber band.

Data Table 1

Answers will vary. But the amount of CO_2 produced should increase with time, except for flask 1, which should produce no CO_2 gas. The amount of CO_2 produced should increase from flask 1 to flask 3.

Part 1: Determining the effects of sugar concentration on CO_2 production

1. Use a wax pencil to label nine 250 mL Erlenmeyer flasks or glass bottles 1 to 9.

2. Add 5 g of dried yeast to each flask.

3. Add 1 g of sucrose (table sugar) to flask 2; add 5 g of sucrose to flask 3.

4. To each flask, add 200 mL of lukewarm water (37°C). Gently swirl each flask to mix the contents.

5. Place a deflated spherical balloon over the mouth of each flask, and secure it tightly with a rubber band.

6. Place all the flasks in a culture incubator set at 37°C to maintain the initial temperature. If an incubator is not available, wrap the flask with an insulating material to prevent heat loss.

7. After 30 minutes, measure the circumference of each balloon. Record your measurements in **Data Table 1.** Measure the balloon's circumference again after 60 minutes and after 90 minutes. Record these measurements in **Data Table 1.**

Data Table 1—*Sugar concentration (g)*

	30 minutes		60 minutes		90 minutes	
	Balloon circum-ference (cm)	CO_2 volume (mL)	Balloon circum-ference (cm)	CO_2 volume (mL)	Balloon circum-ference (cm)	CO_2 volume (mL)
Flask 1: 0 g						
Flask 2: 1 g	11	22	13	37	16	69
Flask 3: 5 g	19	120	23	205	27	330

Part 2: Determining the optimum growth temperature for yeast

8. To each of flasks 4, 5, and 6, add 5 g of sucrose.

9. Add 200 mL of ice water (5°C) to flask 4, 200 mL of room-temperature water (20°C) to flask 5, and 200 mL of lukewarm water (37°C) to flask 6. Gently swirl each flask to mix the contents.

10. Place a deflated spherical balloon over the mouth of each flask, and secure it tightly with a rubber band.

11. Place flask 4 in an ice bath and flask 6 in a culture incubator set at 37°C to maintain the initial temperature. If an incubator is not available, wrap the flask with an insulating material to prevent heat loss.

Data Table 2

Answers will vary. Flask 4 should produce little or no CO_2 gas. The gas volume for flask 6 will be greater than the gas volume for flask 5.

12. After 30 min, measure the circumference of each balloon. Record your measurements in **Data Table 2.** Measure the balloon's circumference again after 60 minutes and after 90 minutes. Record these measurements in **Data Table 2.**

Data Table 2—*Temperature (°C)*

	30 minutes		60 minutes		90 minutes	
	Balloon circum-ference (cm)	CO_2 volume (mL)	Balloon circum-ference (cm)	CO_2 volume (mL)	Balloon circum-ference (cm)	CO_2 volume (mL)
Flask 4: 5°C						
Flask 5: 23°C	15	60	19	120	22	180
Flask 6: 37°C	19	120	23	205	27	330

Part 3: Determining the effect of pH on carbon dioxide production

13. To each of flasks 7, 8, and 9, add 5 g of sucrose.

14. To each of three 250 mL beakers, add 200 mL of lukewarm (37°C) water. Label one beaker "pH 2" and add white vinegar drop by drop while stirring until the pH equals 2. Label a second beaker "pH 6" and add, while stirring, white vinegar drop by drop or Na_2CO_3 in small amounts until the pH equals 6. Label the third beaker "pH 11" and slowly add, small amounts of Na_2CO_3 while stirring, until the pH equals 11.

15. Add the solution in beaker pH 2 to flask 7. Add the solution in beaker pH 6 to flask 8, and add the solution in beaker pH 11 to flask 9.

16. Gently swirl each flask to mix the contents.

17. Place a deflated spherical balloon over the mouth of each flask, and secure it tightly with a rubber band.

Data Table 3

Answers will vary. pH values for Flask 8 and Flask 9 should decrease, and for Flask 7 may decrease or remain the same. Flask 8 will have the highest volume of CO_2 gas.

18. Place flasks in a culture incubator set at $37°C$ to maintain the initial temperature. If an incubator is not available, wrap the flask with an insulating material to prevent heat loss.

19. After 30 min, measure the circumference of each balloon. Record your measurements in **Data Table 3.** Measure the balloon's circumference again after 60 minutes and after 90 minutes. Record these measurements in **Data Table 3.** After 90 minutes, test the solutions in each flask with pH paper. Record the results in **Data Table 3.**

Data Table 3—pH							
	30 minutes		**60 minutes**		**90 minutes**		
	Balloon circumference	**CO$_2$ volume (mL)**	**Balloon circumference**	**CO$_2$ volume (mL)**	**Balloon circumference**	**CO$_2$ volume (mL)**	**Final pH**
Flask 7: pH 12	9	12	11	22	14	46	2
Flask 8: pH 6	19	120	23	205	27	330	4.5
Flask 9: pH 11	10	17	12	29	13	37	9

Cleanup and Disposal

Disposal

Solutions containing yeast and sucrose may be poured down the drain.

20. Clean all apparatus and your lab station. Return equipment to its proper place. Dispose of chemicals and solutions in the containers designated by your teacher. Do not pour any chemicals down the drain or put them in the trash unless your teacher directs you to do so. Wash your hands thoroughly after all work is finished and before you leave the lab.

CALCULATIONS

Organizing Data

1. Calculate the volume of the CO_2 gas in each balloon by substituting each circumference measurement in **Data Tables 1–3** into the following equations. Record your volume calculations in the appropriate tables to the correct number of significant figures.

$$r = \frac{circumference}{2\pi}$$

$$V = \frac{4}{3}\pi r^3$$

Answers will vary. In general, volume should increase from flask 1 to flask 3, increase from flask 4 to flask 6, and increase from flask 7 to flask 8, then decrease from flask 8 to flask 9. Check that calculations are performed correctly.

1. Analyzing Results How is carbon dioxide production dependent on sugar concentration?

As the amount of sugar increases, the amount of CO_2 produced

increases. When no sugar is present, no CO_2 is produced.

2. Analyzing Results Based on your results in Part 2 of the investigation, how does temperature affect CO_2 gas production?

The most CO_2 is produced at 37°C. Less CO_2 is produced at room

temperature, and none is produced at 5°C. Therefore, warmer

temperatures favor fermentation.

3. Analyzing Results Compare the beginning pH values for each flask in Part 3 of this investigation with the final pH values. Explain any variation.

As sugar is consumed, the CO_2 produced reacts with water mole-

cules to form H_2CO_3, causing the pH to decrease with time. A pH

value of 2 is a bit too acidic, so there should be little or no CO_2

production.

4. Evaluating Data Based on your results in Part 3, does baker's yeast thrive in an environment that is acidic or basic? Justify your answer.

Most types of yeast thrive in an acidic environment of pH 4 to

6.0. In a highly acidic environment (pH = 2), CO_2 production is

greatly decreased or stops altogether. Similarly, a strongly basic

environment (pH = 11), impedes the production of CO_2.

**GENERAL
CONCLUSIONS**

1. **Predicting Outcomes** What would happen to carbon dioxide production if the fermentation reaction were allowed to run overnight?

Carbon dioxide production would peak and then stop when the

sugar is depleted because sugar is required for the reaction to

take place.

2. **Applying Conclusions** Carbon dioxide bubbles are responsible for the spongy texture of yeast breads. Based on your lab results, suggest favorable conditions for preparing yeast bread.

Yeasts thrive in lukewarm temperature, slightly acidic pH, and

with a sufficient sugar source.

3. **Designing Experiments** Design an experiment that tests the effect of other sugars or sugar substitutes on carbon dioxide production by yeast. Determine the extreme temperature range that baker's yeast can withstand.

Answers will vary; Sugar substitutes or nonnutritive sweeteners

should not react with the yeast.

4. **Analyzing Methods and Designing Experiments** Warmth favors the fermentation process. Describe how the procedure in Part 2 of this experiment could be modified to determine the *ideal* temperature for CO_2 production.

Answers will vary, but procedures should show experiments done

over a range of temperatures from 0°C to 50°C. Above 50°C, the

yeast cells die.

Name_____

Date_____ Class_____

Solutions: Rock Formations

OBJECTIVES

Recommended time:
60 min; then observe
after 24 h

- **Prepare** a supersaturated salt solution.
- **Observe** salt crystal formation.
- **Relate** the formation of salt deposits to rock formation.

INTRODUCTION

Solution/Material Preparation

1. You may want to assign this experiment to be done at home if there is not enough classroom space for the setups to sit overnight undisturbed.

When ground water containing $CaCO_3$ seeps through the cracks and pores of a cave ceiling, it drips downward toward the floor of the cave. Exposure to the air causes the water to evaporate, leaving $CaCO_3$ deposits on the ceiling and floor of the cave. These deposits are called stalactites and stalagmites, respectively. In this experiment, you will simulate the formation of stalagmites and stalactites by evaporating a supersaturated solution of Epsom salts.

SAFETY

Required Precautions

- Read all safety precautions, and discuss them with your students.
- Safety goggles and a lab apron must be worn at all times.
- In case of a spill, use a dampened cloth or paper towels to mop up the spill. Then rinse the cloth in running water at the sink, wring it out until it is only damp, and put it in the trash.

  Always wear safety goggles and a lab apron to protect your eyes and clothing. If you get a chemical in your eyes, immediately flush the chemical out at the eyewash station while calling to your teacher. Know the location of the emergency lab shower and the eyewash station and the procedure for using them.

 Do not touch any chemicals. If you get a chemical on your skin or clothing, wash the chemical off at the sink while calling to your teacher. Make sure you carefully read the labels and follow the precautions on all containers of chemicals that you use. If there are no precautions stated on the label, ask your teacher what precautions you should follow. Do not taste any chemicals or items used in the laboratory. Never return leftovers to their original containers; take only small amounts to avoid wasting supplies.

 Call your teacher in the event of a spill. Spills should be cleaned up promptly according to your teacher's directions.

 Never put broken glass in a regular waste container. Broken glass should be disposed of properly.

MATERIALS

- 110 g Epsom salts, $MgSO_4{\cdot}10H_2O$
- 1 Al foil, 20 cm × 20 cm
- 250 mL hot water
- 150 mL beakers, 2
- 400 mL beaker
- cotton string, unwaxed, 30 cm
- large paper clips

PROCEDURE

Pre-lab Discussion

Students should be familiar with concentration terms and realize that a solute can be recovered by evaporating the solvent.

Disposal

● Mix the Epsom salt and solutions together. Dissolve the salt. Adjust the pH to 7 with dilute HCl or NaOH, then pour the solution down the drain. If preferred, the salt can be recovered and reused. Mix the salt and solution in a labeled wide-mouth jar. When the water has evaporated, affix the lid to the jar, and store the jar on a shelf until the next time the experiment is done.

1. Add the 110 g of Epsom salts to 250 mL of **HOT** tap water. Stir.

2. Attach a large paper clip to each end of the 30 cm piece of string.

3. Soak the string in the Epsom salt solution for 5 minutes.

4. While the string is soaking, lay the Al foil on the tabletop. Set the two 150 mL beakers at opposite edges of the foil.

5. Remove the string from the Epsom salt solution. Place one end of the string in each of the 150 mL beakers.

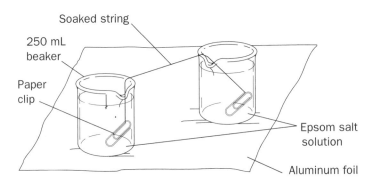

6. Fill each 150 mL beaker with 125 mL of the Epsom salt solution, and let sit for 30 minutes. Observe the string, and record your observations. Let the setup stand overnight, and observe the results.

Observations after 30 minutes:

The center of the string should be crusty and may be thicker than the rest of the string.

Observations after 24 hours:

The center of the string should be heavy with salt deposit. The area of the foil directly beneath the salt deposit should also contain a small mound of salt.

Cleanup and Disposal

7. Clean all apparatus and your lab station. Return equipment to its proper place. Dispose of chemicals and solutions in the containers designated by your teacher. Do not pour any chemicals down the drain or put them in the trash unless your teacher directs you to do so. Wash your hands thoroughly after all work is finished and before you leave the lab.

QUESTIONS

1. Evaluating Methods Explain the purpose of soaking the string and tell why the ends of the string are left in the beaker.

Soaking the string decreases the time needed to observe salt

formation because the experimenter will not need to wait for the

solution to be wicked up. The ends of the string remain in the

beaker so that the salt solution continues to flow toward the salt

deposit.

2. Inferring Results How would the concentration of the salt solution be affected if cold water was used instead of hot water?

The solution would be less concentrated.

3. Inferring Results Would salt deposits form if the solution were less concentrated? Explain your reasoning.

Yes; formation and buildup would be slower because more water

would have to evaporate before a deposit could be observed.

GENERAL CONCLUSIONS

1. Analyzing Results Using your observations of the salt deposits, suggest an explanation for stalagmite formation.

If evaporation is slow enough, the water drips to the ground and

deposits form on the floor rather than on the ceiling of the cave.

2. **Evaluating Methods** If the paper clips were not attached to the ends of the string, how would your experimental results be affected?

The weight of the deposit formation would cause the string to

sag until the ends of the string were pulled out of the beakers. A

mound would form instead of stalactites.

3. **Applying Concepts and Designing Experiments** Water picks up ions as it percolates through the ground. The types of ions present and their concentrations change as the depth of the water changes. Halite, NaCl, and sylvite, KCl, are two minerals that form when water evaporates near the surface of the earth. Halite, or rock salt, has the same composition as table salt, but containers of rock salt are labeled as poison because the salt contains impurities. Explain how the rock salt could be purified.

Answers will vary but should suggest that rock salt could be

recrystalized from solution.

4. **Designing Experiments** Halite and sylvite are found together in salt brines. If these are the only two salts present in a brine solution, design an experiment that would separate the two salts from each other.

Answers will vary but could include fractional distillation, shifting

ion concentrations to favor one salt over another, or recrystaliz-

ing from alcohol.

HOLT
ChemFile
L A B   P R O G R A M

EXPERIMENT **D13**

A Close Look at Soaps and Detergents

OBJECTIVES

Recommended time:
60 minutes

- **Observe** the action of surfactants in reducing surface tension.
- **Evaluate** the foaming capacity of commercial soaps and synthetic detergents.
- **Determine** the efficiency of soaps and synthetic detergents in emulsifying a fat sample in water.

INTRODUCTION

Solution/Material Preparation

1. To prepare a bar-soap solution, add a single soap bar to 1 L of distilled water. Stir the solution occasionally, and allow the mixture to stand overnight. Remove the remainder of the bar. The mixture can be used directly. Alternatively, add 0.15 g of the soap bar to 10 mL of distilled water.
2. To prepare a liquid-soap or detergent solution, add 150 mL liquid soap (detergent) to 1 L distilled water. The mixture is a 1.5% (v/v) solution and can be used directly.

Soap is a surface-active agent (surfactant) that lowers the surface tension of water, allowing fat or oil-bearing soil particles to be suspended in water (emulsification). Although soaps are excellent cleansers in soft water, they are ineffective in hard-water conditions. Hard water contains aqueous salts of magnesium, calcium, and iron. When soap is used in hard water, the insoluble calcium salts of the fatty acids and other precipitates are deposited as curds. This precipitate is commonly referred to as bathtub ring or scale. To overcome this problem, inexpensive synthetic detergents were developed around 1950. Although immensely popular, these compounds were not biodegradable and were replaced in the mid-1960s.

Synthetic detergents, like soaps, have a long hydrocarbon chain labeled as a tail that is attached to a hydrophilic head. Synthetic surfactants are classified as anionic, cationic, or nonionic, depending on the type of hydrophilic head. The hydrophilic portion of a synthetic detergent is often a sulfonate, phosphate, or something other than the carboxyl group that is found in the soap. Therefore, synthetic surfactants are effective in both hard and soft water. No precipitates form when calcium, magnesium, or iron ions are present in solution. The most widely used group of synthetic detergents is the linear alkyl sulfonates (LAS). LAS are straight-chain compounds that have 10 or more carbon atoms and are easily degraded by bacteria.

Because water is a polar substance, it cannot remove dirt from fabrics if the dirt is suspended in oil and grease, which are nonpolar. The hydrocarbon tail of a detergent molecule dissolves itself in an oily substance but leaves the polar head outside the oily surface. Many detergent molecules continue to orient themselves in this way until the dirt containing oil is encapsulated. The oil droplet is then lifted away from the fabric and suspended in the water as a droplet or micelle.

SAFETY

3. To prepare a solid detergent solution, mix 15 g of powdered detergent with 1 L of distilled water. The mixture is a 1.5% (w/v) solution and can be used directly.

Always wear safety goggles and a lab apron to protect your eyes and clothing. If you get a chemical in your eyes, immediately flush the chemical out at the eyewash station while calling to your teacher. Know the location of the emergency lab shower and the eyewash station and the procedures for using them.

4. To prepare a 4% $CaCO_3$ solution, add 40 g of $CaCO_3$ to 800 mL distilled water; stir until all the salt dissolves, and add distilled water to a final volume of 1 L.

5. To prepare a 0.5% alcoholic Sudan IV stain, add 0.5 g of Sudan IV (lipid and fat stain) to 10 mL of denatured iso-propyl alcohol.

 Do not touch any chemicals. If you get a chemical on your skin or clothing, wash the chemical off at the sink while calling to your teacher. Make sure you carefully read the labels and follow the precautions on all containers of chemicals that you use. If there are no precautions stated on the label, ask your teacher what precautions you should follow. Do not taste any chemicals or items used in the laboratory. Never return leftovers to their original containers; take only small amounts to avoid wasting supplies.

 Call your teacher in the event of a spill. Spills should be cleaned up promptly, according to your teacher's directions.

 Never put broken glass in a regular waste container. Broken glass should be disposed of properly.

Never stir with a thermometer because the glass around the bulb is fragile and might break.

MATERIALS

- 0.5% alcohol solution of Sudan IV
- 4% $CaCO_3$ solution
- 1.5% detergent solution
- detergent (powdered and liquid)
- lard
- soap (bar and liquid)
- 500 mL beaker
- 125 mL Erlenmeyer flask
- balance
- fabric, two 2-in. squares
- hot plate
- medicine dropper
- metric ruler
- No. 5 solid rubber stopper
- stirring rod
- thermometer
- wax paper
- wax pencil

PROCEDURE

6. To prepare the colored fat sample, melt 4 to 5 tablespoons of lard in a double boiler. Once the lard is melted, add enough Sudan IV dye to obtain a dark pink or light red color. Solidify the fat by refrigerating it overnight. The dye will let students observe what happens to the fat as it is mixed with water and detergent.

7. Lard (hog fat) can be obtained from a grocery store; keep it refrigerated.

8. Cut 2 in. squares of cotton fabric.

Part 1: Investigating the action of a surface-acting agent (surfactant)

1. Using a medicine dropper, carefully add single drops of distilled water to a square of clean fabric. Describe what occurs.

> The water droplets bead up on the fabric's surface; very little
>
> water penetrates the fabric.

2. Using a second medicine dropper, carefully add drops of detergent solution on top of the beaded water drops. Describe what happens.

> The detergent reduced the surface tension and allowed the water
>
> to penetrate rapidly into the fabric.

Procedural Tips
- Fabrics other than cotton may be used. You may wish to compare surfactant interaction with natural and synthetic fabrics.
- Any detergent/soap solution can be used in **Part 1.**
- Choose at least 5 representative soap and detergent brands in liquid, powder, and bar forms. Be sure to choose brands that list ingredients.
- All soaps and detergents will foam in soft (distilled) water. Soaps but not detergents foam in 4% $CaCO_3$.

Techniques to Illustrate
- Show students how to correctly measure the foam level. Emphasize that the measurement begins at the interface between the liquid and the foam.

Part 2: Evaluating cleansing characteristics of soaps and synthetic detergents

3. Use a wax pencil to label a set of five 5 Erlenmeyer flasks 1, 2, 3, 4, and 5.

4. Obtain five soap solutions from your teacher. Record in the **Data Table** the manufacturer's brand name next to each sample number. Then add 10 mL of each soap or detergent sample to its respective numbered flask.

5. Stopper flask 1 securely with a rubber stopper. Hold the flask with your thumb on the stopper, and shake it vigorously for 15 seconds. Allow the solution to stand for 30 seconds. Observe and measure the level of the foam. Record this value in the **Data Table.** Repeat this process for flasks 2 through 5. Which product has the highest foam level?

   Answers will vary, depending on the brands tested.

   _____

6. To each flask add 4 drops of 4% $CaCO_3$ solution from a medicine dropper. Stopper the flask, and with your thumb on the stopper, shake vigorously for 15 seconds. Allow the solution to stand for 30 seconds. Observe and measure the level of foam. Record this value in the **Data Table.** Describe how $CaCO_3$ affects a surfactant's ability to foam.

   Ions from $CaCO_3$ (hard-water conditions) inhibit the foaming

   action in soaps. Calcium deposits (curds) or film deposits on the

   sides of the flask may be observed with some soaps. Detergent

   foaming is unaffected by hard-water conditions; precipitate de-

   position and film should not be noted.

Part 3: Evaluating the effectiveness of soaps and synthetic detergents in emulsifying fat

7. Fill a 500 mL beaker with 250 mL of distilled water. Set the beaker on a hot plate and heat the water to 55°C. Try to maintain this temperature as closely as possible throughout the experiment.

8. Measure a 0.1 g sample of colored fat onto a piece of wax paper.

9. Add the fat sample to the water in the beaker. Use a glass stirring rod to swirl the water and fat. Describe what happens.

   The lard (fat) forms a distinct globular layer on the surface of the

   water.

   _____

10. Fill a graduated cylinder with exactly 100 mL of test sample 1.

Pre-Lab Discussion

● Discuss emulsification of the fat by the soap or detergent sample. Students should understand that in **Part 3** the fat changes from a solid globular blob into tiny globules, eventually dispersing throughout the water.
● Review the calculation of efficiency for this experiment.
● Discuss why foaming is important to the cleansing action of surfactants.
● Students should understand the relationship between the terms: *surfactant, soap,* and *synthetic detergent.*

11. While stirring constantly, slowly add sample 1 to the water/lard mixture until the fat is completely emulsified, as evidenced by the dispersion of the colored fat in water. Record in the **Data Table** the volume of detergent necessary to completely emulsify the lard (fat) globule.

12. Identify each sample tested as either a detergent or a soap in the **Data Table.** Based upon emulsification data, which soap and synthetic detergent brand is the most effective emulsifier? Give reasons for your answer.

The most effective emulsifiers require the least number of mL to

emulsify the fat.

Cleanup and Disposal

13. Clean all apparatus and your lab station. Return equipment to its proper place. Dispose of chemicals and solutions in the containers designated by your teacher. Do not pour any chemicals down the drain or put them in the trash unless your teacher directs you to do so. Wash your hands thoroughly after all work is finished and before you leave the lab.

Data Table 1—Comparing Soap and Detergent Brands

Sample number	Foam level in water (cm)	Foam level in 4% CaCO₃ (cm)	Volume needed to emulsify 0.1 g of fat (mL)	Efficiency rating (g/mL)	Soap or detergent
1 Softsoap™	2.3	1.3	100+	0.001 estimated	soap
2 Dial™	1.7	1.1	100+	0.001 estimated	soap
3 Wisk™ Free and Clear	1.2	1.2	100+	0.001 estimated	detergent
4 Arm & Hammer™ Heavy Duty	1.1	1.2	100+	0.001 estimated	detergent
5 Joy™ dish-washing liquid	2.0	1.2	60	0.0017	detergent
6 Cascade™ powder	0.1	0.1	52	0.0019	detergent

CALCULATIONS

• Point out that liquid detergents are better because they contain a larger amount of non-ionic surfactants than their dry counterparts. Liquid detergents are very effective at removing oily-type soils by emulsification. Anionic surfactants need help from other ingredients to prevent partial inactivation by water hardness ions.

Disposal

Unused solutions can be stored for reuse. Waste solutions can be washed down the drain.

1. **Organizing Data** For each sample tested, calculate its relative efficiency as a dirt and grease remover by dividing the 0.1 g of fat emulsified by the volume in milliliters of detergent added to the fat/water mixture.

$$\frac{0.1 \text{ g lard (fat)}}{35 \text{ mL soap sample}} = 0.028 \text{ g fat/mL}$$

2. **Organizing Data** Construct a bar graph comparing the efficiency of soaps and detergents in emulsifying fat. Place surfactant samples along the x-axis and the efficiency along the y-axis.

1. Applying Conclusions You are demonstrating the effect of hard water on sudsing. You have all the demonstration materials except a source of hard water. All that is available to you is milk or a roll of antacid tablets. Can the demonstration proceed? Justify your answer.

Yes. Milk contains Ca^{2+} ions. Certain antacids contain $CaCO_3$. As

the mixture is stirred, the fatty acids in the soapy solution are

converted into their insoluble calcium salts.

2. Analyzing Data and Inferring Conclusions According to entries in the **Data Table,** which would be more efficient at dirt removal: a liquid detergent or a solid detergent? Explain.

Liquid detergents; efficiency values, and possibly foam heights,

should be higher.

GENERAL CONCLUSIONS

1. Inferring Conclusions Many product labels on detergents state that the product will not harm septic systems. Suggest a reason for this based on your foam data.

The product will not produce calcium curds that can add scale

and plug plumbing.

2. Inferring Conclusions Visit your local grocer and look for a popular brand of colorless hand gel. Write down the ingredients and compare them with the ingredients in various shampoos. Summarize your conclusions.

Formulations (major ingredients) are almost identical.

Name_____

Date_____ Class_____

HOLT
ChemFile
LAB PROGRAM

Acids and Bases: Lemon Cheese

NOTE: This experiment should be done at home or in the home economics lab at school.

OBJECTIVES

Recommended time:
2 h

- **Observe** the effects of acid on proteins.
- **Demonstrate** proficiency in determining the pH of a solution.
- **Filter** a mixture gravimetrically.

INTRODUCTION

Milk contains proteins. These proteins remain dispersed in the normal pH range of milk, which is 6.3 to 6.6. If the milk is acidified, the proteins clump together. The milk separates into a solid mass and a light-colored liquid. Heat makes this curdling effect more severe. In this activity, you will prepare a soft cheese by adding lemon juice to warmed milk and separating the curd from the whey.

MATERIALS

- 2 L whole milk
- 6.0 mL lemon juice
- pH papers
- salt (optional)
- herbs (optional)
- 2 qt saucepan
- colander
- cooking (candy) thermometer
- cheesecloth
- spatula or knife
- spoon
- stove or hot plate

PROCEDURE

Techniques to Demonstrate

• Students may be unfamiliar with the method of filtering with cheesecloth as a funnel. You can illustrate this process for them.

1. Pour the milk into the saucepan. Test the pH of the milk with pH paper. Record the results in the Data Table.

2. Slowly heat the milk until its temperature is about 40–45°C. Stir the milk frequently while heating.

    Never stir with a thermometer because the glass around the bulb is fragile and might break.

3. Turn off the heat, and remove the saucepan from the stove.

4. Measure out the lemon juice. Test the pH of the juice with pH paper. Record the results in the Data Table.

5. Pour the juice into the milk while stirring. Test the pH of the mixture with pH paper. Record the results.

6. Let the milk mixture stand for 15 minutes. Record the appearance in the Data Table.

7. Line a colander with cheesecloth. The edges of the cloth should hang over the sides of the colander.

Procedural Tips

● This lab was designed as a home lab assignment. However, if your school has a home economics department, you may want to arrange to use the kitchen facility. If the laboratory classroom is used, do not allow students to eat the cheese; instead, place the cheese in a transparent plastic bag, seal the bag, label it "DO NOT EAT," and place it in the trash can.

Pre-lab Discussion

● Students should be familiar with the structure of proteins and with the pH scale.
● If the thermometer that students use has only a Fahrenheit scale, they can use the conversion equation, $°F = \frac{9}{5}°C + 32$. In this case, $75°C = 167°F$.

QUESTIONS

Disposal

See **Procedural Tips** for disposal of the cheese when the lab is performed in the laboratory classroom. If the lab is performed in a sanitary kitchen, then the cheese may be eaten, and no special disposal requirements are needed.

8. Pour the contents of the saucepan into the colander.

9. Pull together the corners of the cheesecloth to form a sack, and allow the contents to drain for about 1 hour.

10. Open the sack, and scrape the cheese from the cloth into a bowl. Add salt and herbs to taste.

Cleanup and Disposal

 12. Clean your work area and all apparatus. Return equipment to its proper place. Wash your hands thoroughly after all work is finished.

Data Table

	Milk	Lemon juice	Mixture
pH	6.3–6.6	around 4.0	varies

Observations from Step 6:

Accept all reasonable answers that indicate formation of a solid and a thinning of the liquid.

1. Inferring Conclusions Consider the pH readings shown in the Data Table. What is the function of the lemon juice?

lowers the pH of the milk

2. Analyzing Results How does the appearance of the mixture from step 6 compare with the appearance of the milk before it was heated?

Accept all reasonable answers. Students should recognize that the milk has separated into a solid and a liquid and that the color and relative clarity of the liquid have changed.

3. Applying Ideas Consider the pH of the lemon juice. Could vinegar be used instead of lemon juice?

yes

EXPERIMENT **D15**

How Effective Are Antacids

OBJECTIVES

Recommended time:
45 minutes

- **Compare** the neutralization ability of antacids.
- **Infer** which antacid(s) tested is most effective.

INTRODUCTION

Solution/Material Preparation

• Wear safety goggles, a face shield, impermeable gloves, and an apron when you prepare the HCl and NaOH solutions. Work in a chemical fume hood known to be in operating condition, and have another person stand by to call for help in case of an emergency. Be sure you are within a 30 s walk from a safety shower and eyewash station known to be in good operating condition.

An acidic stomach is necessary for good health, but excessive stomach acid can produce acute discomfort or contribute to ulcers. An effective antacid neutralizes just enough excess acid to alleviate pain and discomfort; it does not bring stomach acids to neutrality (pH 7.0).

The bases most widely used as active ingredients in antacids are of two types: absorbable and nonabsorbable. Absorbable antacids include $NaHCO_3$ and $CaCO_3$ and are very effective at increasing gastric pH. Because they are easily absorbed into the blood, these compounds can also raise blood pH to dangerous levels, causing kidney damage. Nonabsorbable antacids are relatively insoluble salts of weak bases, such as $Al(OH)_3$ and $Mg(OH)_2$. They interact with HCl, forming nonabsorbed or poorly absorbed salts while increasing gastric pH.

Acid secretion is frequently divided into free acid and combined acid. The amount of free acid is determined by titrating gastric secretions to a pH of 3.5. After this titration is performed, the same gastric secretions are titrated to a pH of 8.5; this measures the combined acid. Gastric secretions mixed with food show little or no free acid but a large amount of combined acid. Conversely, when the stomach secretes large quantities of gastric juice while it is almost empty of food, the larger portion of acid is free acid. In this experiment, you will add HCl to antacid samples and then neutralize the excess HCl by titrating to a pH of 3.5 with NaOH.

SAFETY

1. To prepare 1 L of 0.5 M HCl solution, observe the required safety precautions. While stirring, slowly add 41 mL of 12 M HCl to 500 mL of distilled water. Dilute to 1 L.
2. To prepare 1 L of 0.5 M NaOH solution, observe the required safety precautions. While stirring, slowly add 20 g of NaOH to 800 mL of distilled water. Stir to dissolve the solid. Once dissolved, add distilled water to a final volume of 1L.

  Always wear safety goggles and a lab apron to protect your eyes and clothing. If you get a chemical in your eyes, immediately flush the chemical out at the eyewash station while calling to your teacher. Know the location of the emergency lab shower and the eyewash station and the procedures for using them.

 Do not touch any chemicals. If you get a chemical on your skin or clothing, wash the chemical off at the sink while calling to your teacher. Make sure you carefully read the labels and follow the precautions on all containers of chemicals that you use. If there are no precautions stated on the label, ask your teacher what precautions you should follow. Do not taste any chemicals or items used in the laboratory. Never return leftovers to their original containers; take only small amounts to avoid wasting supplies.

 Call your teacher in the event of a spill. Spills should be cleaned up promptly, according to your teacher's directions.

Never put broken glass in a regular waste container. Broken glass should be disposed of properly.

MATERIALS

3. Select 5 antacid brands; include both liquids and tablets. Set containers in a general access area. Use one tablet or 1 tsp (5 mL) per sample.

- 0.5 M HCl
- 0.5 M NaOH
- antacids (tablet and liquid)
- 50 mL buret
- 10 mL graduated cylinder
- beakers, 150 mL and 250 mL
- buret clamp

- mortar and pestle
- pH meter or bromthymol blue indicator
- ring stand
- microspatula
- stirring rod
- wax pencil

PROCEDURE

Required Precautions

- Read all safety precautions, and discuss them with your students.
- Safety goggles and an apron must be worn at all times.
- In case of an acid or a base spill, first dilute with water. Then mop up the spill with wet cloths or a wet cloth mop while wearing disposable plastic gloves. Designate separate cloths or mops for acid and base spills.

Techniques to Demonstrate

- Show students the proper method for filling a buret. Strongly caution students against reaching above their head to fill a buret.
- Demonstrate the end point of bromthymol blue or the use of a pH meter.
- Review how to read a meniscus and how to clean a buret.

Procedural Tips

- If indicator is used with an artificially colored antacid tablet, have students add additional indicator to produce a noticeable color.

Part 1: Computing the milliequivalents of acid neutralized by antacids

1. Use a wax pencil to label a set of five beakers 1, 2, 3, 4, and 5.

2. Obtain five antacid samples from your teacher. Record the manufacturer's brand name and the active ingredients in the **Data Table.** Then prepare one sample per beaker.

For antacid tablets: Crush one tablet, using a mortar and pestle. Using a microspatula, transfer the contents to the appropriately numbered beaker. To the crushed tablet, add 100 mL of 0.5 M HCl. Use a stirring rod to help dissolve the crushed tablet.

For liquid antacids: Pour 5 mL of the antacid into a graduated cylinder. Add 0.5 M HCl to bring the final volume to 100 mL. Then pour this mixture into the appropriately labeled beaker. Use a stirring rod to mix.

Allow the mixtures to stand for 15 minutes. Stir occasionally.

3. While the mixtures are standing, prepare the titration apparatus. Attach a buret clamp to a ring stand. Use a wax pencil to label the buret 0.5 M NaOH. Insert the buret in the buret clamp.

4. Label a 250 mL beaker "Waste." Fill a second 250 mL beaker with approximately 125 mL of 0.5 M NaOH. Carefully pour 5 mL of the 0.5 M NaOH from the beaker into the buret. Rinse the walls of the buret thoroughly with this solution. Allow the solution to drain through the stopcock into the waste beaker. Close the stopcock. Rinse the buret two more times in this manner, using a new 5 mL portion of NaOH solution each time.

5. Fill the buret above the zero mark with 0.5 M NaOH. Place the waste beaker under the buret, and withdraw enough solution to remove any air from the buret tip and to bring the liquid level within the graduated region of the buret. Record the initial volume of NaOH in the **Data Table.**

6. If you are using a pH meter, calibrate the pH probe according to your teacher's instructions. Then insert the pH probe into the beaker containing the first antacid sample to be analyzed. If you are using bromthymol blue indicator, add 10 drops of the indicator to the acid/antacid mixture. Gently swirl the beaker to mix.

Pre-Lab Discussion

• Discuss the back-titration process. Students should understand why HCl is added in excess and why NaOH is the titrant. They should also understand how to determine the end point of a titration.

• Students should understand why the acid/antacid mixtures are allowed to stand for 15 minutes.

7. Place the beaker from step 6 under the buret, and add NaOH dropwise until a pH reading of 3.5 is reached. Record the final buret reading in the **Data Table.** If using a pH meter, remove the pH probe. Rinse it with distilled water before placing it in the next sample.

8. Repeat steps 4 through 7 for the remaining antacid samples.

Cleanup and Disposal

9. Clean all apparatus and your lab station. Return equipment to its proper place. Dispose of chemicals and solutions in the containers designated by your teacher. Do not pour any chemicals down the drain or put them in the trash unless your teacher directs you to do so. Wash your hands thoroughly after all work is finished and before you leave the lab.

CALCULATIONS

• Discuss the difference between free acid and combined acid. Students should be aware that they are titrating free acid rather than combined acid.

1. Organizing Data The FDA requires manufacturers to rate the effectiveness of their antacid product in terms of the amount of acid neutralized by the base. The amount of acid neutralized is expressed in milliequivalents (mEq). Calculate the mEq of acid neutralized by each antacid using the following equation:

$$\text{mEq of acid neutralized} = [V_{\text{HCl used}} \times C_{\text{HCl}}] - [V_{\text{NaOH added}} \times C_{\text{NaOH}}]$$

If 20 mL of base is added, then the mEq of acid neutralized is:
[(30 mL HCl) × (0.5 M HCl)] − [(20 mL NaOH) × (0.5 M NaOH)] = 15 − 10 = 5 mEq of acid neutralized.

Data Table—*Comparing the Neutralization Ability of Liquid and Solid Antacids*					
Sample number	**Buret readings [mL] 0.5 M NaOH** initial final	**Volume of NaOH used to neutralize acid [mL]**	**mEq of acid neutralized**	**Manufacturer's brand name**	**Manufacturer's active ingredient(s)**
1		13–15		Mylanta liquid	200 mg Al(OH)$_3$; 200 mg Mg(OH)$_2$
2		29–34		Maalox Extra Strength liquid	500 Al(OH)$_3$; 450 Mg(OH)$_2$
3		8–10		Tums tablet	500 mg CaCO$_3$
4		5–7		Di-Gel liquid	280 mg CaCO$_3$; 28 mg Mg(OH)$_2$
5		13–15		Rolaids tablet	550 mg CaCO$_3$; 110 mg Mg(OH)$_2$

Disposal

Set out three disposal containers for the students: one for unused acid solutions, one for unused base solutions, and one for partially neutralized substances and the contents of the waste beaker. One at a time, slowly combine solutions while stirring. Adjust the pH of the final waste liquid with 1 M acid or base until the pH is between 5 and 9. Pour the neutralized liquid down the drain.

QUESTIONS

GENERAL CONCLUSIONS

2. **Organizing Data** Prepare a bar graph that summarizes the milliequivalents of acid neutralized by brand sample. Along the horizontal axis, indicate the brand name. Along the vertical axis, record the milliequivalents of acid neutralized either per tablet or per 5 mL of liquid.

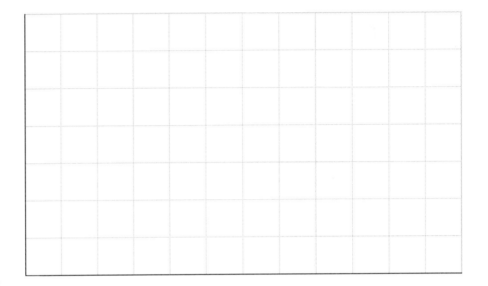

3. **Applying Data** Based on the graph summarizing the milliequivalents of acid neutralized by an antacid tablet or liquid dose, which brand is most effective? Support your answer.

Answers will vary depending on the brands tested. The brand with the largest milliequivalents per tablet or 5 mL dose should be identified.

1. **Analyzing Data** Using data from the **Data Table,** are the liquid forms of popular antacids more effective than tablets in neutralizing acid?

Generally, the data suggest that this relationship does exist.

1. **Analyzing Information** Copy the advertising claims made on the antacid brands evaluated. Are claims of "extra strength" supported by your data? Does this experiment allow you to evaluate claims for "quicker acting"? Support your answers.

Claims of extra strength should be supported by data because these products contain larger amounts of active ingredients. No, time to dissolve or react were not examined, so the data cannot be used to evaluate claims of quick-acting formulas.

EXPERIMENT

Titration of Aspirin

OBJECTIVES

Recommended time:
1 lab period

- **Conduct** an acid-base titration using an aspirin tablet as the acid.
- **Measure** the volume of a known concentration of a basic solution needed to reach the end point.
- **Calculate** the number of moles of base needed to reach the end point.
- **Calculate** the molar mass of the aspirin.

INTRODUCTION

Required Precautions

- Read all safety precautions, and discuss them with your students.
- Safety goggles and a lab apron must be worn at all times.
- In case of a spill, use a dampened cloth or paper towels to mop up the spill. Then rinse the cloth in running water at the sink, wring it out until it is only damp, and put it in the trash.
- Broken glass should be disposed of in a clearly labeled box lined with a plastic trash bag. When the box is full, close it, seal it with packaging tape, and set it next to the trash can for disposal.
- Students should avoid direct contact with the NaOH solution. If contact with this solution occurs, the affected areas should be thoroughly rinsed with water.

Aspirin is slightly acidic and reacts with bases in neutralization reactions. If the reaction is followed with a visual indicator, a color change will occur when the acid is neutralized. The number of moles of base consumed and the number of moles of acid in the sample can be calculated from the volume of the base needed to obtain the color change. The relationship between these two mole values is determined from the balanced chemical equation for the reaction.

In this experiment, you will titrate an aspirin tablet with a known concentration of a sodium hydroxide solution. The end point will be determined with phenolphthalein or with thymol blue. The following chemical equation describes the double-replacement reaction that will be observed in this experiment.

$$\text{Aspirin–H} + \text{NaOH} \longrightarrow \text{Aspirin–Na} + \text{H}_2\text{O}$$
$$\text{(acid)} \qquad \text{(base)}$$

Because the mole ratio between the two reactants is 1:1, the number of moles of the base (NaOH) equals the number of moles of the acid (Aspirin-H) at the titration's end point. The number of moles of NaOH is obtained by rearranging the following equation.

$$\text{molarity of NaOH} = \frac{\text{amount of NaOH (in mol)}}{\text{volume of NaOH (in L)}}$$

$$\text{moles of NaOH} = (\text{molarity of NaOH}) \times (\text{volume of NaOH})$$

The molar mass of solid aspirin can be determined with the following formula.

$$\text{molar mass} = \frac{\text{mass (in g)}}{\text{amount of aspirin (in mol)}}$$

The mass of the aspirin inside the tablet is 325 mg. Inactive ingredients, such as binders, are added to the aspirin during the manufacturing process. Therefore, the actual mass of the tablet exceeds 325 mg because it is not 100% pure aspirin.

SAFETY

Always wear safety goggles and a lab apron to protect your eyes and clothing. If you get a chemical in your eyes, immediately flush the chemical out at the eyewash station

while calling to your teacher. Know the locations of the emergency lab shower
and the eyewash station and the procedures for using them.

**Do not touch any chemicals. If you get a chemical on your skin or
clothing, wash the chemical off at the sink while calling to your
teacher.** Make sure you carefully read the labels and follow the precautions on all containers of chemicals that you use. If there are no precautions stated
on the label, ask your teacher what precautions you should follow. Do not taste
any chemicals or items used in the laboratory. Never return leftovers to their original containers; take only small amounts to avoid wasting supplies.

Call your teacher in the event of a spill. Spills should be cleaned up
promptly, according to your teacher's directions.

Never put broken glass in a regular waste container. Broken glass
should be disposed of properly.

MATERIALS

- 100 mL distilled water in a wash bottle
- 50 mL 0.100 M NaOH
- 40 mL 95% ethanol
- phenolphthalein or thymol blue indicator, 6–8 drops
- 325 mg nonbuffered aspirin tablets, 2
- 100 mL graduated cylinder
- 250 mL Erlenmeyer flasks, 2
- buret
- buret clamp
- funnel, short stem, for filling buret
- ring stand

PROCEDURE

1. Assemble the ring stand, buret clamp, and buret.

2. Fill the buret with 0.100 M NaOH solution. The bottom of the meniscus
   should rest on the 0.00 mL line.

3. Place a 325 mg aspirin tablet in an Erlenmeyer flask. Add about 20 mL of
   distilled water. Wait for 20–30 seconds. The tablet will break apart as it ab-
   sorbs water and swells. Then add about 20 mL of ethyl alcohol to dissolve
   the aspirin.

4. Add 3–4 drops of the assigned indicator solution. *If you use phenolphthalein, the color will change from colorless to faint pink. If you use thymol blue, the color will change from faint yellow to faint blue.*

5. Titrate to the end point with 0.100 M NaOH. Record in the data table the
   volume of base that was required.

6. Repeat steps 1–5 with a second aspirin tablet.

- The end point is more readily observed with thymol blue because the color change (from yellow to light blue) is more distinct than with phenolphthalein.
- Generic aspirin tablets give very reproducible results.
- The volume of the 0.100 M NaOH solution that is needed for each trial is 18–20 mL.

Cleanup and Disposal

7. Clean all apparatus and your lab station. Return equipment to its proper place. Dispose of chemicals and solutions in the containers designated by your teacher. Do not pour any chemicals down the drain or put them in the trash unless your teacher directs you to do so. Wash your hands thoroughly after all work is finished and before you leave the lab.

Data Table

Aspirin mass (g)	NaOH conc. (M)	NaOH volume (mL)	NaOH volume (L)
0.325	0.100	18.00	0.01800
0.325	0.100	18.20	0.01820

CALCULATIONS

Pre-Lab Discussion

- Review the concept of molarity and how to do calculations involving mole conversions and mole ratios in a chemical equation. The balanced word equation can be used to illustrate that the mole ratio between the starting materials is 1:1. Point out that the number of moles of aspirin is determined indirectly from the number of moles of NaOH consumed.
- Students might be tempted to obtain the mass of the whole aspirin tablet. Remind students that the aspirin tablet contains 325 mg of the aspirin plus inactive ingredients. So the total mass of the tablet is greater than 325 mg. None of the inactive ingredients in the aspirin tablet react with the NaOH.
- Students should be allowed to struggle with the calculations in this experiment. You may want to discuss the theory behind the experiment as a post-lab activity instead of a pre-lab activity.

1. Organizing Ideas The label on the bottle of aspirin states that 325 mg of aspirin is present in each tablet. Convert this mass to grams, and add this value to the data table.

325 mg = 0.325 g

2. Organizing Ideas What is the mole ratio between the two starting materials in this chemical reaction?

1:1

3. Applying Ideas Convert the volumes of sodium hydroxide from milliliters to liters. Add these values to your data table in the appropriate spaces.

The volumes should be divided by 1000 because 1000 mL = 1 L.

4. Applying Ideas Calculate the number of moles of sodium hydroxide present in each trial if the concentration of the NaOH was 0.100 M.

0.0180 L NaOH × 0.100 M NaOH = 0.00180 mol NaOH

Disposal
• The liquid wastes generated in this experiment can be poured down the sink. Any excess NaOH solution can be saved for other classes or for additional experiments.

QUESTIONS

5. **Applying Ideas** According to the mole ratio, how many moles of the aspirin were present in each of your trials?

$$0.00180 \text{ mol NaOH} \times \frac{1 \text{ mol aspirin}}{1 \text{ mol NaOH}} = 0.00180 \text{ mol aspirin}$$

6. **Applying Ideas** Use your two answers in the previous problem to calculate the molar mass of aspirin for each trial.

$$0.00180 \text{ mol aspirin} = \frac{0.325 \text{ g aspirin}}{X}$$

$X = 181$ amu (The actual molar mass of aspirin is 180.)

7. **Organizing Ideas** What is the purpose of the indicator in an acid-base titration?

The indicator provides an observable visible change at the end

point of a titration. It tells you when an end point has been

reached.

_____

_____

_____

8. **Evaluating Methods** How could the insoluble white binder affect the detection of the end point in this experiment?

The white binder could mask the end point, since we are looking

for slight color changes in the indicator.

_____

_____

EXPERIMENT **D17**

Household Indicators

OBJECTIVES

Recommended time:
90 minutes (45 minutes
for Parts 1 and 2 and
45 minutes for Part 3)

- **Extract** anthocyanins from red cabbage leaves.
- **Prepare** a pH indicator paper with the red cabbage anthocyanins.
- **Construct** a color indicator chart for anthocyanins from red cabbage.
- **Compare and evaluate** the accuracy of indicator papers for recording pH values of common items.

INTRODUCTION

Solution/Material Preparation

Wear safety goggles, a face shield, impermeable gloves, and an apron when you prepare the HCl or acetic acid solutions. Work in a chemical fume hood known to be in operating condition, and have another person stand by to call for help in case of an emergency. Be sure you are within a 30 second walk from a safety shower and eyewash station known to be in good operating condition.
1. To prepare 1 L of 0.1 M HCl solution, observe the required safety precautions. While stirring, slowly add 8.3 mL of 12 M HCl to 500 mL of distilled water. Dilute to 1 L. Pour solution into dropper bottles for lab use.

A visual indicator is a chemical substance that reflects the nature of the chemical system in which it is placed by changing color. Most visual indicators are complex organic molecules that exist in multiple colored forms, one of which could be colorless, depending on the chemical environment. Many visual indicators are used to test a solution's acidity.

Acid-base indicators respond to hydronium ion concentrations, $[H_3O^+]$. Acidic solutions have an excess of H_3O^+ ions, while basic or alkaline solutions have few H_3O^+ ions. A measure of the $[H_3O^+]$ is pH. Chemists use p to mean "power," so pH means the power of the hydronium ions. Formally, pH is defined as the negative logarithm of the $[H_3O^+]$ of a solution.

The most common acid-base indicator is litmus, a blue coloring matter extracted from various species of lichens. The chief component of litmus is azolitmin. Widely distributed among the higher plants, anthocyanins constitute most of the yellow, red, and blue colors in flowers and fruits. Anthocyanins are excellent acid-base indicators because they exhibit color changes over a wide range of pH values. Red cabbage is a ready source of this pigment. Unlike litmus and anthocyanins, universal indicator is a mixture of various synthetic indicator molecules. Universal indicator provides pH scale coverage from 1 to 14 pH units.

In this investigation, you will extract the pigment from red-cabbage leaves and use it to prepare strips of pH indicator paper. You will use these paper strips to test the pH of common household materials. Then you will compare your results with pH tests made with litmus and universal pH paper.

SAFETY

Always wear safety goggles and a lab apron to protect your eyes and clothing. If you get a chemical in your eyes, immediately flush the chemical out at the eyewash station while calling to your teacher. Know the location of the emergency lab shower and the eyewash station and the procedures for using them.

2. To prepare 1 L of 0.1 M CH₃COOH solution, observe the required safety precautions. Add 5.8 mL of glacial acetic acid to 500 mL of distilled water. Dilute to 1 L. Pour the solution into dropper bottles for lab use.

3. To prepare 1 L of 0.1 M H₃BO₃ solution, observe the required safety precautions. Add 6.2 g of H₃BO₃ to 500 mL of distilled water. Dilute to 1 L. Pour the solution into dropper bottles for lab use.

4. To prepare 1 L of NaHCO₃ solution, observe the required safety precautions. Add 8.4 g of NaHCO₃ to 500 mL distilled water. Dilute to 1 L. Pour the solution into dropper bottles for lab use.

MATERIALS

5. To prepare 1 L of 0.1% ammonia solution, observe the required safety precautions. Add 100 mL of household ammonia to 900 mL of distilled water. Pour the solution into dropper bottles for lab use.

6. To prepare 1 L of NaOH solution, observe the required safety precautions. Add 4 g of NaOH to 500 mL of distilled water. Dilute to 1 L. Pour the solution into dropper bottles for lab use.

7. Select at least 10 of the common household substances listed in **Data Table 2**. Make sure "squeezings" and "slices" are made fresh, at the start of the laboratory. pH readings can change due to oxidation.

8. Set up a clothesline so that students can dry their filter paper circles. Penciled initials allow students to identify their papers.

 Do not touch any chemicals. If you get a chemical on your skin or clothing, wash the chemical off at the sink while calling to your teacher. Make sure you carefully read the labels and follow the precautions on all containers of chemicals that you use. If there are no precautions stated on the label, ask your teacher what precautions you should follow. Do not taste any chemicals or items used in the laboratory. Never return leftovers to their original containers; take only small amounts to avoid wasting supplies.

 Call your teacher in the event of a spill. Spills should be cleaned up promptly, according to your teacher's directions.

 Never put broken glass in a regular waste container. Broken glass should be disposed of properly.

 Do not heat glassware that is broken, chipped, or cracked. Use tongs or a hot mitt to handle heated glassware and other equipment because hot glassware does not always look hot.

- 0.1 M CH₃COOH
- 0.1% household ammonia
- 0.1 M H₃BO₃
- 0.1 M HCl
- 0.1 M NaHCO₃
- 0.1 M NaOH
- household materials:
    ammonia, apple, distilled water, fresh egg, grapefruit, fruit jelly, ginger ale, lemon, milk, milk of magnesia, molasses, orange, mineral water, sauerkraut, sweet potato, tomato, salt water, and tap water
- 250 mL beakers
- clothespins
- colored pencils (red, rose, purple, blue, green, yellow)
- dropper bottles for acid solutions
- 9 cm diameter filter paper
- forceps
- hot mitt or pot holder
- hot plate
- litmus paper, neutral
- medicine dropper
- metric ruler
- red cabbage
- fine-point scissors
- 6-well spot plate, white
- string
- universal pH paper, (1–14 pH units)
- wax pencil

PROCEDURE

Part 1: Extracting anthocyanins from red cabbage

1. Choose a red-cabbage leaf that has a dark purple color. Tear the leaf into small pieces.

2. Fill a 250 mL beaker 2/3 full with the leaf pieces. Add enough distilled water to just cover the pieces of cabbage leaf. Place the beaker on a hot plate, and bring the water to a slow boil. Continue heating for 5 minutes. Then turn off the heat, and allow the mixture to cool for 10 to 15 minutes.

Part 2: Dyeing indicator strips

3. Using a hot mitt, remove the beaker from the hot plate. Carefully decant the cooled purple liquid into a second 250 mL beaker. Dispose of the cabbage-leaf material as directed by your teacher.

4. Using a pencil, write your initials on six filter paper circles. Using forceps, submerge the filter paper into the beaker containing the cabbage extract. Make sure that the papers are thoroughly wet. Then, using forceps, remove the filter papers from the beaker and allow them to dry thoroughly.

5. After drying, use scissors to cut from the filter paper 30 individual strips, each measuring 1 cm × 6 cm.

Part 3: Constructing a color indicator chart for the anthocyanin-dyed pH paper

6. Use a wax pencil to label six wells in a spot plate 1 to 6.

7. Following the table below, add 10 drops of each named acid or base solution to its labeled well.

1	2	3	4	5	6
0.1 M CH$_3$COOH	0.1% household ammonia	0.1 M H$_3$BO$_3$	0.1 M HCl	0.1 M NaHCO$_3$	0.1 M NaOH

8. Into each of the six wells, dip a separate strip of the indicator paper prepared in **Part 2**. Record the color in **Data Table 1**. Rinse off the well plates.

Data Table 1		
pH	Substance	Indicator color
1.0	0.1 M hydrochloric acid, HCl	red
2.9	0.1 M acetic acid, CH$_3$COOH	rose
5.2	0.1 M boric acid, H$_3$BO$_3$	purple
8.4	0.1 M sodium bicarbonate, NaHCO$_3$	blue
11.1	0.1 M ammonia, NH$_3$	green
13.0	0.1 M sodium hydroxide, NaOH	yellow

9. Use colored pencils and the entries in **Data Table 1** to fill in the following color indicator chart.

pH color chart for anthocyanins extracted from red cabbage

Red		Rose		Purple			Blue		Green			Yellow	
1	2	3	4	5	6	7	8	9	10	11	12	13	14

Part 4: Comparing and evaluating the accuracy of indicator papers

10. Test each item listed in **Data Table 2** with universal pH paper, litmus paper, and the anthocyanin paper prepared in **Part 2**. Test each of the liquids listed by placing 10 drops of the liquid in a spot-plate well. Test the juice of each solid. Record the pH values in **Data Table 2.**

Data Table 2

Common substance	Measured pH	pH (neutral litmus paper)	pH (anthocyanin dyed paper)	pH (universal pH paper)
Ginger ale	2.0–4.0	<7	3–4	4.0
Lemon slice (lemon juice)	2.2–2.4	<7	1–2	2.0
Apple cider	2.4–2.9	<7	3–4	2.0
Apple slice	2.9–3.3	<7	4–5	2.0
Grapefruit (cut)	3.0–3.3	<7	2–3	2.0
Jellies, fruit	3.0–3.3	<7	2–4	2.0
Orange slice	3.0–4.0	<7	1–2	2.0
Sauerkraut	3.4–3.6	<7	1–2	2.0
Tomato slice	4.1–4.4	<7	5–7	2–4
Molasses	5.0–5.4	<7	6–8	5.0
Sweet potato slice	5.3–5.6	<7	5–7	5.0
Mineral water	6.2–9.4	<7	5–7	5.0
Milk	6.4–6.8	<7	5–7	5–6
Tap water	6.5–8.0	>7	8–9	8–9
Distilled water (CO$_2$-free)	7.0	7	5–7	8–9
Fresh egg	7.6–8.0	>7	9–11	7.0
Salt water	8.0–8.4	>7	8–9	7–8
Milk of Magnesia	10.5	>7	10–12	10–11
Household ammonia	10.5–11.9	>7	9–10	10.0

Cleanup and Disposal

11. Clean all apparatus and your lab station. Return equipment to its proper place. Dispose of chemicals and solutions in the containers designated by your teacher. Do not pour any chemicals down the drain or put them in the trash unless your teacher directs you to do so. Wash your hands thoroughly after all work is finished and before you leave the lab.

QUESTIONS

Disposal

Have students discard boiled cabbage leaves in a plastic bag that can be tied, labeled "Not for Consumption," and disposed of with the trash. Unused solutions can be stored for later use. Used solutions can be poured down the drain with lots of water.

1. **Organizing Conclusions** Do all three pH test papers cover the entire pH scale range? Use your data to support your answer.

No. Neutral litmus paper has a pH range of 4.5 (red) to 8.5

(blue). Universal pH paper and cabbage indicator papers have

wider pH scale ranges—essentially 1–14.

2. **Analyzing Data** Which pH test paper was the most accurate within its respective pH scale range? Explain your reasoning.

Answers will vary. The Universal pH test paper is the most accu-

rate. Both the cabbage indicator and litmus papers provide a

general indication of pH.

3. **Inferring Conclusions** Suggest reasons for the variation in measured pH values.

Answers will vary but may include pH scale differences, complex-

ity of indicator molecules, concentration of the indicator in the

test paper or of the test substance on the paper, variation in

color perception, or testing technique.

4. **Analyzing Data** Use the pH of a fresh egg to determine the hydronium ion concentration for the egg.

Because the pH of a fresh egg is about 8, the $[H_3O^+]$

is 0.000 000 01 M.

$pH = 8$
$-\log [H_3O^+] = 8$
$\log [H_3O^+] = -8$
$[H_3O^+] = -10^{-8}$
$= 0.000 000 01$

GENERAL CONCLUSIONS

1. **Inferring Conclusions** The pH of cow's milk is approximately 6.5. When milk spoils, we sometimes say it has gone sour because of the lemon-like taste it develops. From your knowledge of food pH chemistry, what is happening to the milk's pH?

Lemon has an acidic pH and a sour taste. As acidity increases, the

pH drops, so one can assume that the taste is due to an increase

in acidity and a decrease in pH. (This acidity comes from the for-

mation of lactic acid by milk-inhabiting bacteria, *Lactobacillus aci-*

dophilus.)

2. **Designing Experiments** The acid-base indicator, phenolphthalein, is also a mild laxative and can be the active ingredient in commercial chocolate-flavored laxatives. Phenolphthalein is not soluble in water, but it is soluble in rubbing alcohol (70% isopropyl alcohol). How would you evaluate its effectiveness as an indicator?

Answers will vary. First, the indicator has to be extracted in a

small volume of rubbing alcohol. Its color range should be

evaluated in a method similar to this experimental procedure.

(Phenolphthalein is colorless to pH 8.5 and pink to deep red

above pH 9.)

ChemFile

EXPERIMENT **D18**

Shampoo Chemistry

OBJECTIVES

Recommended time:
45 minutes

- **Compare** selected shampoos for the following characteristics: pH, foam volume and retention, oil dispersal, and viscosity.
- **Identify** shampoo ingredients.

INTRODUCTION

Solution/Material Preparation

1. Set out containers of shampoo.

Required Precautions

- Read all safety precautions, and discuss them with your students.
- Safety goggles and an apron must be worn at all times.

Procedural Tips

- Obtain various clear shampoo brands for student analysis. Include baby shampoos and shampoos that are also conditioners. Use multiple product containers so students can easily review ingredient listings.
- If time or shampoo supplies are short, have each student study three shampoo samples instead of five.

Hair is a lifeless structure composed of the cross-linked polymer protein keratin. Cross-linking gives hair its strength. A strand of hair has a central core (cortex) that contains its coloring pigments. A thin, scaly sheath, the cuticle, encases the cortex. Within the scalp, sebaceous glands secrete an oily substance (sebum), which gives hair its gloss, and keeps the scales of the cuticle lying flat. Too much sebum makes the hair feel greasy and dirty. Too little sebum makes the hair dry and unmanageable.

The principle function of a shampoo is to produce lather and remove dirt from the hair and scalp without removing all of the oils. Lather is the foaming created when a surfactant is agitated in water. A good shampoo will have at least twice its volume in lather, or a 2:1 ratio. Lather stabilizers help maintain the foam condition during product use. The longer the foam lasts, the more effective the cleansing action is. Dirt and grease trapped in the foam fraction of the shampoo are more difficult to rinse away than dirt and grease dispersed in the liquid fraction. India ink contains a dispersion of minute carbon particles in water. In this experiment, you will use India ink as a test dirt. By observing whether the tiny carbon particles are trapped in the shampoo's foam, you will evaluate the cleansing effectiveness of the product.

We also expect our shampoos to provide fullness and luster to our hair. The pH of a shampoo is responsible for luster. Acidic pH shampoos tend to tighten the hair cuticle and scales, allowing light to reflect evenly. Basic pH shampoos tend to swell cuticle scales, resulting in light that leaves hair looking "dull" and "flat." Because hair is a protein, it is slightly acidic, so shampoos with pH values between 4.5 and 6.5 produce the best luster or shiniest hair. The body of a shampoo is its viscosity. The higher the viscosity is, the thicker the shampoo—a quality that implies a premium product.

SAFETY

Always wear safety goggles and a lab apron to protect your eyes and clothing. If you get a chemical in your eyes, immediately flush the chemical out at the eyewash station while calling to your teacher. Know the location of the emergency lab shower and the eyewash station and the procedures for using them.

Pre-Lab Discussion

• Discuss the importance of foam, and review calculation of the foam-to-liquid ratio. Make sure students understand why the volume of foam is divided by 25 mL.
• Remind students to keep their thumb over the stopper when shaking the graduated cylinder.

Do not touch any chemicals. If you get a chemical on your skin or clothing, wash the chemical off at the sink while calling to your teacher. Make sure you carefully read the labels and follow the precautions on all containers of chemicals that you use. If there are no precautions stated on the label, ask your teacher what precautions you should follow. Do not taste any chemicals or items used in the laboratory. Never return leftovers to their original containers; take only small amounts to avoid wasting supplies.

Call your teacher in the event of a spill. Spills should be cleaned up promptly, according to your teacher's directions.

Never put broken glass in a regular waste container. Broken glass should be disposed of properly.

MATERIALS

- 150 mL beakers, 5
- 100 mL graduated cylinder
- clear shampoo samples
- India ink
- copper BBs

- medicine dropper
- No. 3 stopper
- stopwatch or watch with a second hand
- universal pH paper
- wax pencil

PROCEDURE

• Remind students that FDA labeling regulations require that the highest concentration ingredient be listed first, followed by the next highest, and so on.

Disposal

Recover the BBs, rinse them thoroughly, dry, and store them in an appropriately labeled container for later use. Unused shampoos and India ink should be stored for use at a later time. Shampoo solutions can be washed down the drain with lots of water.

Part 1: Comparing shampoo characteristics

1. **Preparing a 1% Shampoo Solution** Obtain from your teacher five shampoo samples. Use a wax pencil to label five beakers 1 to 5.

2. Pour 2 mL of a shampoo sample into the graduated cylinder. Fill the cylinder to the 100 mL mark with distilled water. Pour the diluted shampoo into its correspondingly numbered beaker. Repeat the dilution process for the four remaining shampoo samples. Thoroughly rinse the graduated cylinder between shampoo samples.

3. **Testing pH** Using universal pH test paper, determine the pH value of each shampoo sample. Record the values in **Data Table 1.** Which shampoo brand(s) would give hair the best shine or luster?

   Answers will vary. Chosen brand(s) should have pH values that

   are slightly acidic. Basic pH shampoos tend to dull hair.

4. **Testing Lather Volume and Retention** Carefully pour 25 mL of one shampoo sample into a 100 mL graduated cylinder. To avoid producing foam, slowly pour the diluted shampoo sample down the *side* of the graduated cylinder.

5. Firmly stopper the cylinder. Shake the cylinder up and down 10 times. Record the volume of foam in **Data Table 1.** Calculate the ratio of foam to liquid by dividing the volume of foam by 25 mL. Enter this ratio in **Data Table 1.**

6. Remove the stopper from the graduated cylinder. Record in **Data Table 1** the volume of foam remaining after 1 minute and after 5 minutes.

7. Repeat steps 4 through 6 for the remaining shampoo samples. Do all shampoos pass the 2:1 ratio test for lather? Justify your answer.

<u>Answers will vary. Most shampoos should maintain a 2:1 ratio for</u>

<u>at least 1 minute. Check that the data support the justification.</u>

8. **Testing Dispersal** Carefully pour 25 mL of one shampoo sample into a 100 mL graduated cylinder. Avoid foaming. Add one drop of India ink, stopper the cylinder, and shake the cylinder up and down 10 times. Remove the stopper from the graduated cylinder. Determine if the India ink has been dispersed (evenly spread) throughout either the liquid portion or the foam portion of the shampoo sample. Place a check mark next to "liquid" and/or "lather" in **Data Table 1**.

9. **Testing Shampoo Viscosity** Fill a graduated cylinder to the 100 mL mark with an *undiluted* shampoo sample. Release a copper BB at the top of the cylinder. Using a stopwatch or a watch with a second hand, determine the time for the BB to fall to the bottom of the cylinder. The longer the drop time is, the thicker, more viscous, the shampoo. Record the data in **Data Table 1**.

10. Repeat steps 8 and 9 for the other shampoo samples.

Data Table 1—*Shampoo Brand Characteristics*

Sample number and brand	pH	Foam ratio (mL)		Foam retention/ Height of foam (cm)		Ink dispersal (√)	Viscosity [BB drop time] (s)
1 Jergens Moisturizing Body Shampoo	8.0	volume: 80 ratio: 3:2:1		1 min 80	5 min 75	liquid [√] lather [ ]	3
2 Jhirmack Salon Formula Conditioning	6.5	volume: 80 ratio: 3:2:1		1 min 80	5 min 78	liquid [√] lather [ ]	5
3 St. Ives Swiss Formula	7.5	volume: 83 ratio: 3:3:1		1 min 82	5 min 79	liquid [√] lather [ ]	23
4 Revlon Flex Balsam and Protein	7.5	volume: 90 ratio: 3:6:1		1 min 90	5 min 90	liquid [√] lather [ ]	2
5 Thermasilk	6.5	volume: 87 ratio: 3:5:1		1 min 86	5 min 84	liquid [√] lather [ ]	3

Part 2: Performing an "ingredient" audit

11. Examine the ingredients listing for at least one shampoo and conditioner product. Using the **Information Table** as a guide, determine in which functional category each ingredient belongs, and write the name of the ingredient in the matching column of **Data Table 2.**

Information Table—*Shampoo ingredients*

Ingredient	Weight %	Functional category and function
Purified water • Aqua	60	solvent
Triethanolammonium lauryl sulfate • sodium lauryl sulfate • sodium laureth sulfate • ammonium lauryl sulfate • ammonium laureth sulfate • cocamidopropyl betaine • cococamphodiacetate • sodium cocglyceryl ether sulfonate • sodium lauryl sarcosinate • sodium laureth–13	32	surfactants (cleansing agents)
Myristic acid • citric acid	4	pH adjustment (luster)
Cetyl alcohol • stearyl alcohol • caprylic acid • Hydrogenated lanolin • polyethylene glycol • glycol sterate • palmitic acid • PEG-80		thickeners (give texture, appearance, and flow to the product)
Oleyl alcohol • panthenol • amino acids • protein • collagen • dimethicone	2	conditioning agents (give hair a softer, thicker feel)
Cocamide MEA • lauramide MEA • lauric DEA • polysorbate-20	1	lather stabilizers (maintain lather during use)
Fragrance (parfum)	1	perfumes
Guar hydroxypropyltrimonium chloride • dicetydimonium chloride • behentrimonium chloride • behanalkonium betaine • benzalkonium chloride • quaterium-18 • stearalkonium chloride •cetrimonium chloride • polyquaternium 10 • quaterium 15	1	quaternary ammonium compounds (hair detanglers)
Formaldehyde • methyl paraben • propylparaben • phenoxyethanol • quateternium 15 • DMDM hydantoin	0.5	preservatives (anti-microbials)
Glycerin • sorbitol • glycols • mucopolysaccharides • hyaluronic acid • sodium PCA • glycoshingolipids • sorbitan laurate	0.5	humectants (hold water in the hair shaft, giving hair bounce and fullness)
Polyphosphates • EDTA	0.5	sequestrants and chelating agents (to soften hard water)
FDC blue, red, yellow	0.5	colorants

EXPERIMENT D18 continued

Data Table 2—*Ingredient Audit*

Product Name: Revlon Flex Balsam and Protein

Functional category	Present in ingredients listing (√)	Named ingredients
Solvent	√	water
Surfactant(s)	√	ammonium lauryl sulfate, sodium C14-17 alkyl SEC sulfonate, sodium laureth sulfate, cocamidepropyl betaine
pH adjustment	√	citric acid
Thickeners	√	methyl cellulose, hydroxyl propyl methyl cellulose
Conditioning agents	√	panthenol
Lather stabilizers	√	lauramide DEA
Perfumes	√	fragrance
Preservatives	√	methyl paraben, propyl paraben
Humectants		
Sequestrant(s) and chelating agents	√	tetrasodium EDTA
Colorants	√	FD&C yellow #5, D&C red #33, FD&C red #4, FD&C yellow #6

Cleanup and Disposal

12. Clean all apparatus and your lab station. Return equipment to its proper place. Dispose of chemicals and solutions in the containers designated by your teacher. Do not pour any chemicals down the drain or put them in the trash unless your teacher directs you to do so. Wash your hands thoroughly after all work is finished and before you leave the lab.

QUESTIONS

1. **Analyzing Data** According to entries in the **Information Table,** which ingredients in a shampoo play no part in the removal of oils and dirt?

pH adjustment chemicals, conditioning agents, perfumes, quaternary ammonium compounds, preservatives, and colorants play no part in dirt and oil removal. Surfactants are the primary cleansing agents supported by lather stabilizers. Sequestrants and chelating agents allow the surfactants to work in hard-water conditions.

2. **Analyzing Data** Which shampoo brand(s) have the greatest viscosity? Do these same brands bear advertising that promotes extra body?

Answers will vary. BB drop time should be about equal except for brands that advertise extra body.

GENERAL CONCLUSIONS

1. **Inferring Conclusions** What do you think is meant by the advertising term "Natural pH?"

a pH that is slightly acidic

2. **Analyzing Information** Which shampoo brand provides the best value? Give reasons for your answer.

Answers will vary. Justifications may include inexpensiveness, pH promotes luster, least amount of dirt in foam, or best foaming action. Accept all reasonable answers.

3. **Inferring Conclusions** Examine the **Information Table** entries. Then suggest differences in the composition of shampoos and conditioners.

Conditioners will contain a larger number of thickeners, humectants, and conditioners than shampoos; conversely, conditioners tend to have fewer cleansing agents.

Name_____

Date_____ Class_____

EXPERIMENT

Rust Race

OBJECTIVES

Recommended time:
Day 1: 60 minutes to set up investigations
Day 4: 45 minutes to observe and record results

- **Test** metals for corrosion.
- **Measure** oxygen consumption during metal corrosion.
- **Compare** the effectiveness of rust inhibitors.

INTRODUCTION

Solution/Material Preparation

1. To prepare 1 L of a 2% salt solution, add 20 g of NaCl to 1 L of tap water.
2. To prepare 1 L of a vinegar solution, mix 250 mL of vinegar with 750 mL of tap water.
3. To prepare bleach solution for washing the steel wool, mix 1 part bleach to 9 parts water. Be sure to observe proper safety precautions when using bleach.

When a metal corrodes, it oxidizes, or transfers electrons to a nonmetal, such as oxygen. Copper reacts with oxygen in the air to form the soft green substance called verdigris. When iron is exposed to moist air, it reacts with oxygen to form the reddish brown substance iron oxide, or rust.

Other factors such as temperature, salt, acidity, and air or water pollutants, also affect the rate of metal corrosion. For example, iron and steel tend to corrode much more quickly when exposed to salt, such as that used to melt snow, or to a moist, salty environment, such as areas near the ocean. This is because dissolved salts increase the conductivity of the aqueous solution formed at the surface of the metal.

A metal can also be protected from corrosion by being coated with paint, rust inhibitors, or special films. These coatings prevent exposure of the metal to oxygen or moisture. Galvanizing is the most widely used method to protect iron and steel products from corrosion. During this process, a thin coating of zinc is applied to a metal. When the galvanized metal is exposed to elements in the atmosphere, the zinc coating oxidizes instead of the iron or steel.

In this experiment, you will test metals for oxidation, measure oxygen consumption during metal corrosion, and identify ways to protect iron metal from corrosion.

MATERIALS

4. Obtain any commercially available rust inhibitor from a hardware store.

- bleach solution
- clear nail polish
- galvanized nails, 3
- iron nails, 8
- paper clips, 3
- pennies, 3
- quarters, 3
- petroleum jelly
- rust inhibitor, commercially available
- 2% salt solution
- steel wool, plain

- vinegar solution
- 100 mL beakers, 5
- 500 mL beaker
- 50 mL graduated cylinder
- balance
- forceps
- metric ruler
- shredded paper
- test tubes (16 × 125 mm), 2
- water
- wax pencil

SAFETY

Always wear safety goggles and a lab apron to protect your eyes and clothing. If you get a chemical in your eyes, immediately flush the chemical out at the eyewash station while calling to your teacher. Know the location of the emergency lab shower and the eyewash station and the procedures for using them.

Do not touch any chemicals. If you get a chemical on your skin or clothing, wash the chemical off at the sink while calling to your teacher. Make sure you carefully read the labels and follow the precautions on all containers of chemicals that you use. If there are no precautions stated on the label, ask your teacher what precautions you should follow. Do not taste any chemicals or items used in the laboratory. Never return leftovers to their original containers; take only small amounts to avoid wasting supplies.

 Call your teacher in the event of a spill. Spills should be cleaned up promptly, according to your teacher's directions.

 Never put broken glass in a regular waste container. Broken glass should be disposed of properly.

PROCEDURE

Part 1: Testing metals for corrosion

1. Use a wax pencil to label three 100 mL beakers "Water," "Salt," and "Vinegar." Using a graduated cylinder, pour into each beaker 50 mL of the solution corresponding to the beaker's label.

2. Using forceps, place the following in each beaker: a penny, a quarter, an iron nail, a paper clip, and a galvanized nail. Then place the three beakers in an undisturbed area of the laboratory, and let them stand undisturbed for 4 or 5 days.

3. After 4 or 5 days, use forceps to remove your test materials from each beaker, laying the contents of each beaker on separate paper towels. Carefully observe each object for any physical changes that may have taken place. Record any signs of metal corrosion (loss of shine, change in color, rust or verdigris formation) in **Data Table 1.**

Part 2: Measuring oxygen consumption during metal corrosion

4. Use a wax pencil to label two test tubes "Steel Wool" and "Control."

5. Determine the mass of 1 g of steel wool to the nearest 0.1 g. Rinse the steel wool with bleach to remove any rust inhibitors it may contain. This ensures that oxygen can access the surface of the metal for oxidation to take place.

6. Pull apart the strands of the steel wool so that they are loosely packed and fluffy. Fill the bottom third of the test tube labeled **"Steel Wool"** with the loosely packed steel wool. The steel wool should make contact with the sides of the test tube and should not slide out when the tube is inverted.

7. Measure the mass of 1 g of shredded paper to the nearest 0.1 g and fill the bottom third of the test tube labeled **"Control."** The paper should make contact with the sides of the test tube and should not slide out when the tube is inverted.

8. Pour about 200 mL of water in a 500 mL beaker. Invert the test tubes from steps 6 and 7 and place them open end down in the beaker. The test tubes should rest against the side of the beaker. Keep the tubes' open ends fully beneath the water level.

Data Table 1—*Testing Metals for Corrosion*

Observations of beaker with:

Test material	Water			Salt solution			Vinegar solution		
	Shine	Color	Rust	Shine	Color	Rust	Shine	Color	Rust
Iron nail	none	reddish brown	yes	none	reddish brown	yes	none	reddish brown	yes
Galavanized nail	dulls	silverish gray	no	dulls	silverish gray	no	dulls	silverish gray	no
Paper clip	dulls	silverish gray	no	dulls	silverish gray	no	dulls	silverish gray	no
Quarter	dulls	silverish gray	no	dulls	silverish gray	no	dulls	silverish gray	no
Penny	lost luster	copper/ greenish	yes	none	greenish	yes	none	greenish	yes

Disposal

Set up a disposal container for any sharp objects such as nails and other metal objects. Waste vinegar and salt solutions may be poured down the drain. Dispose of all other waste materials according to the directions provided by the manufacturer of the product.

9. Use a wax pencil to mark the water level in each test tube. Use a metric ruler to measure the height of water in each tube. Also measure the height of the test tube above the water. Record your measurements in **Data Table 2.** Leave your setup undisturbed for 24 hours.

10. After 24 hours, mark the new water level in each test tube. Use a ruler to measure the new water level in each test tube. Record your results in **Data Table 2.** Why did the water rise in the test tube with the steel wool but not in the test tube with water?

    The water rises to fill the space left behind when oxygen molecules in the air above the water combine with the steel wool to form rust. No reaction takes place between the paper and oxygen, so the water does not rise.

11. Using forceps, remove the steel wool from the test tube and determine its mass. Record your results in **Data Table 2.** Using forceps, remove the shredded paper from the test tube and determine its mass. Record your results in **Data Table 2.** Carefully observe the steel wool strands and the shredded paper. Describe any corrosion that has taken place.

    Observations:

    Answers will vary.

Data Table 2—*Measuring Oxygen Consumption*

Measurement	Test tube with steel wool	Test tube with shredded paper
Height of test tube above water level (cm)	7.5 cm	7.5 cm
Height of water at start (cm)	5 cm	5 cm
Height of water after 24 h (cm)	6.5 cm	same as initial height
Water level increase (cm)	1.5 cm	0
Percent increase in water level	20%	0%
Percent of oxygen consumed by volume	20%	0%
Initial mass (g)	1.0	1.0
Final mass (g)	1.1	1.0
Change in mass (g)	0.1	0.0

NOTE: Height of the water and test tube would vary depending on the setup used.

Part 3: Protecting metals from corrosion

12. Obtain 5 iron nails. Thoroughly coat the entire surface of the first nail with petroleum jelly, coat the second nail with a commercially available rust inhibitor, coat the third nail with clear nail polish, and coat the fourth nail with a coating of your choice. Leave the fifth nail uncoated; it is your control.

13. Pour 50 mL of salt solution in a 100 mL beaker. Using forceps, place your coated nails in the beaker. Be sure the nails do not touch each other. Place the beaker containing the nails in an undisturbed area of the laboratory for 4 or 5 days.

14. After 4 or 5 days, use forceps to remove the nails from the beaker. Lay the nails on separate paper towels. Carefully observe each nail for any physical changes that may have taken place. Record any signs of metal corrosion (loss of shine, change in color, or rust formation) in **Data Table 3.**

Data Table 3—*Protecting Metals from Corrosion*

Test material	Observations		
	Shine	Color	Rust
Iron nail (control)	lost luster	reddish brown	yes
Iron nail coated with petroleum jelly	N/A	original color, exposed areas are reddish brown	no
Iron nail coated with clear nail polish	none	same as above	yes, some
Iron nail coated with rust inhibitor	original luster	original color	no
Iron nail with coating of your choice	Answers will vary, depending on the coating.		

Cleanup and Disposal

15. Clean all apparatus and your lab station. Return equipment to its proper place. Dispose of chemicals and solutions in the containers designated by your teacher. Do not pour any chemicals down the drain or put them in the trash unless your teacher directs you to do so. Wash your hands thoroughly after all work is finished and before you leave the lab.

CALCULATIONS

1. **Organizing Data** Calculate the water level increase for each test tube by subtracting the initial height of the water from the height of the water after 24 hours. Record your results in **Data Table 2.**

2. **Organizing Data** Calculate for both test tubes the percentage increase in water level by dividing the change in water level by the height of the test tube above the water, then multiply the answer by 100. Record your results in **Data Table 2.**

3. **Organizing Data** Calculate the change in the mass of the steel wool by subtracting the initial mass of the steel wool from its final mass. Record your result in **Data Table 2.**

   Students should note a mass increase of 0.1 g or more, depending on the extent of corrosion that has taken place.

4. **Organizing Data** Calculate the change in the mass of the shredded paper by subtracting the initial mass of the paper from its final mass. Record your result in **Data Table 2.**

   Students should note a mass increase of 0.0 g.

5. **Inferring Conclusions** Based on your calculations in item 2, what is the percentage of oxygen consumed by volume in each test tube?

   The water in the tube will rise by the same percentage as the

   oxygen consumed. Because oxygen makes up about 20% of the

   composition of air, students should observe a maximum of 20%

   increase in water level.

1. **Analyzing Data** According to your observations in **Data Table 1,** which solution(s) caused the most corrosion in the materials tested?

   salt and/or vinegar solutions; students' data should support their

   responses.

2. **Analyzing Data** Explain why the steel wool gained mass in **Part 2** of the investigation.

   The steel wool combines with oxygen from the air and produces

   iron oxide (rust).

GENERAL CONCLUSIONS

1. **Designing Experiments** The tarnishing of silver is an oxidation reaction between silver and sulfur in the air or in certain foods. Identify a food that contains sulfur, and design an experiment to demonstrate the tarnishing of silver.

   Answers will vary. One possibility is to push part of a shiny silver

   coin into the egg white of a hard-boiled egg. Remove the coin

   after 1 hour. Because the egg contains sulfur, it will react with sil-

   ver to form a tarnish on the portion of the coin covered by the

   egg white.

2. **Designing Experiments** Design an experiment to determine which brand of steel wool resists rusting the longest.

   Answers will vary. One possible solution is to place 1 g samples of

   each steel wool to be tested in separate test tubes. Add 10 mL of

   vinegar to each test tube. Let the setups sit undisturbed. Observe

   the samples periodically, and record the extent of corrosion in

   each case.

EXPERIMENT **D20**

Measuring the Iron Content of Cereals

OBJECTIVES

Recommended time:
Part 1, 30 minutes;
Part 2, 15 minutes

- **Isolate** and **measure** the iron content in breakfast cereals.
- **Compare** the iron content listed on various cereal boxes to lab results.
- **Observe** the action of acids on iron.

INTRODUCTION

Solution/Material Preparation

- Wear safety goggles, a face shield, impermeable gloves, and an apron when you prepare the HCl or acetic acid solutions. Work under a chemical fume hood known to be in operating condition, and have another person stand by to call for help in case of an emergency. Be sure you are within a 30 s walk from a safety shower and eyewash station known to be in good operating condition.
1. Prepare a 0.2% solution of HCl by diluting 0.2mL of concentrated HCl to a final volume of 100 mL of water.
2. Choose cereals containing differing amounts of iron. Such as the following: Total, Multi-Grain Cheerios, Oatmeal Squares, and Product 19.

Required Precautions

- Read all safety precautions, and discuss them with your students.
- Safety goggles and a lab apron must be worn at all times.

All foods contain at least one of six basic nutrients essential for organisms to grow and function properly: carbohydrates, proteins, lipids, vitamins, minerals, and water. Few foods contain all six nutrients, and most foods contain a concentration of only one or two of these nutrients. But it is important that the foods we eat contain a combination of all six nutrients, so foods are fortified with additional vitamins and minerals. This is one way to ensure a properly balanced diet and to prevent nutritional deficiency diseases.

Flour and products made with flour were first fortified with iron in 1940. The human body requires iron for many functions. Most importantly, iron is used in the production of hemoglobin molecules in red blood cells. The iron contained in hemoglobin attracts oxygen molecules, allowing the blood cells to carry the iron throughout the body. Because red blood cells are being constantly replaced, there is a continuous need for iron in the diet. While an iron overdose is harmful and can cause kidney damage, iron deficiency can lead to anemia, a condition in which red blood cells cannot carry a sufficient amount of oxygen to cells. Iron is found in foods such as liver and other red meats, raisins, dried fruits, whole-grain cereals, legumes, and oysters.

The nutritional content in packaged foods is listed on the "Nutrition Facts" food label found on most food packages. The label provides a listing of all the nutrients, along with their recommended dietary allowances (RDA) based on a 2000 Cal diet. The RDAs are the average dietary amounts of various nutrients that, in the opinion of scientists, we should consume to remain healthy. One staple breakfast food is cereal. Most brands provide a minimum of 25% of the recommended dietary allowance for iron. Typically, very tiny particles of pure powdered metallic iron are mixed in the cereal batter along with other additives. When consumed, the iron particles react with the digestive juices in the stomach and change to a form easily absorbed by the human body. Acidic foods can actually increase iron absorption 3–7 fold.

In this investigation, you will isolate and quantify the amount of iron added to three breakfast cereals and compare the amount collected to the amount listed on the food label. In addition, you will simulate how an acidic environment in the stomach aids in the conversion of iron particles to a form easily absorbed in the body.

SAFETY

• In case of a spill, use a dampened cloth or paper towels to mop up the spill. Then rinse the cloth in running water at the sink, wring it out until it is only damp, and put it in the trash.

Techniques to Demonstrate

• Demonstrate the proper method for attaching the bar magnet to the stir rod.

  Always wear safety goggles and a lab apron to protect your eyes and clothing. If you get a chemical in your eyes, immediately flush the chemical out at the eyewash station while calling to your teacher. Know the location of the emergency lab shower and the eyewash station and the procedure for using them.

 Do not touch any chemicals. If you get a chemical on your skin or clothing, wash the chemical off at the sink while calling to your teacher. Make sure you carefully read the labels and follow the precautions on all containers of chemicals that you use. If there are no precautions stated on the label, ask your teacher what precautions you should follow. Do not taste any chemicals or items used in the laboratory. Never return leftovers to their original containers; take only small amounts to avoid wasting supplies.

 Call your teacher in the event of a spill. Spills should be cleaned up promptly, according to your teacher's directions.

 Never put broken glass in a regular waste container. Broken glass should be disposed of properly.

MATERIALS

• 0.2% HCl
• cereals, 3 brands with high iron content
• 500 mL beakers
• balance
• strong bar magnet
• glass stirring rod

• magnetic stirring bar (optional)
• magnifier (optional)
• mortar and pestle
• test tube
• transparent tape
• wax pencil

PROCEDURE

Procedural Tips

• If a mortar and pestle are not available, place the cereal in a plastic bag and crush the flakes with a rolling pin.
• A magnetic stirrer with stir bar may be substituted for the bar magnet. The magnetic stirrer gives better results.
• You may want to vary this activity by mixing the crushed cereal and water in a sealed plastic bag. Place a large magnet on the outer surface of the bag. Iron particles will start to collect inside the bag, near the magnet.

Part 1: Isolating iron from breakfast cereals

1. Obtain a serving quantity for each of three iron-fortified cereals. Read the "Nutrition Facts" food label for the percentage of iron per serving and record the amount for each cereal in the **Data Table.**

2. Use a wax pencil to label three beakers with the brand name of each cereal.

3. Use a mortar and pestle to grind a serving quantity of each cereal brand into a fine powder. Place each powdered cereal in the appropriately labeled beaker. *Remember to clean the mortar and pestle after each use.*

4. Determine the mass of a bar magnet to the nearest milligram and record your results in the **Data Table.**

5. Tape a bar magnet to the end of a glass stirring rod, leaving most of the bar exposed.

6. Pour 250 mL of warm water into one of the three beakers. Continuously stir the powdered flakes with the bar magnet for 10–15 minutes. The longer the cereal solution is stirred, the more tiny iron particles that will precipitate out of the flakes and attach to the magnet. About 30 minutes of stirring gives the best results.

ChemFile

Pre-Lab Discussion

• Point out the location of the "Nutrition Facts" food label on a cereal box. Discuss the content of the label, and make students aware that the numbers on this label are percentages, not masses. The RDA of iron is 18 mg for a 2000-calorie diet.

Disposal

Combine used and unused acid solutions in a large beaker. Adjust the pH with 1 M NaOH to between 4 and 7, then pour the mixture down the drain.

7. Remove the magnet and carefully drain any excess liquid from it. Determine the mass of the magnet and iron filings to the nearest milligram. Record your results in the **Data Table** and on the chalkboard so that average class results can be determined.

8. Place your bar magnet on a paper towel. Use a magnifying lens to observe the tiny slivers of iron attached to the bottom of the magnet. Record your description of the iron slivers below. Then clean the magnet, saving the iron slivers for step 9. Repeat steps 4–7 for the remaining two cereals.

Observations

_____

_____

_____

_____

Part 2: Simulating the absorption of iron in the body

9. Combine in a test tube all of the iron slivers collected from each cereal.

10. Pour 5 mL of 0.2% HCl in the tube, and let the cereal-HCl mixture stand for 10 minutes. Observe any change(s) in the color, shape, or size of the iron slivers.

Cleanup and Disposal

11. Clean all apparatus and your lab station. Return equipment to its proper place. Dispose of chemicals and solutions in the containers designated by your teacher. Do not pour any chemicals down the drain or put them in the trash unless your teacher directs you to do so. Wash your hands thoroughly after all work is finished and before you leave the lab.

Data Table

	Cereal A: Multi-Grain Cheerios		Cereal B: Total		Cereal C: Quaker Oatmeal Squares	
	Your result	Class average	Your result	Class average	Your result	Class average
Initial mass of magnet (g)	20 g		20 g		20 g	
Final mass of magnet (g)	20.016		20.018		20.014	
Mass of iron recovered (mg)	16 mg		13 mg		11 mg	
Percentage of iron on the label	45%		8%		25%	
Actual mass of iron (mg)	18 mg		18 mg		14.4 mg	
Percent error (%)	11.1%		27.8%		23.6%	

CALCULATIONS

1. **Organizing Data** Calculate the amount of iron extracted from each cereal by subtracting the initial mass of the magnet from the final mass of the magnet. Record your answers in the **Data Table.**

   Answers will vary depending on the cereals chosen and students' technique, but should be less than the actual amounts noted on the RDA label.

2. **Organizing Data** Calculate the actual mass of iron in each cereal tested by multiplying the percentage from the box label by 18 mg (the RDA for iron). Record your answers in the **Data Table.**

   Answers will vary depending on cereal chosen. For Cheerios: 45% x 18 mg = 8.1 mg

3. **Organizing Data** To determine the percent error for each iron measurement, subtract the mass of iron recovered from the actual mass of iron. Then divide the difference by the actual mass of iron and multiply by 100%. Record your answers in the **Data Table.**

   For Cheerios: % error = $\dfrac{18 \text{ mg} - 16 \text{ mg}}{18 \text{ mg}} \times 100 = 11.1\%$

4. **Applying Concepts** A food label lists the serving amount of a particular food as providing 35% of the RDA of iron. If the RDA based on a 2000 calorie diet is 18 mg, what amount of iron must a person consume from other foods to get 100% of the recommended dietary allowance?

   35% of 18 mg is 6.3 mg so a person must consume an additional 11.7 mg of iron.

ChemFile

QUESTIONS

1. **Analyzing Data** Examine your percent error entries in the **Data Table.** How closely do your results verify the claims made by the cereal manufacturers? If your percent error is high, suggest reasons for this.

Accept all reasonable answers. Students' results should justify

their answers. Technique and experimental method are two main

sources of error.

2. **Inferring Conclusions** How do you think the iron slivers isolated from cereal compares with the iron found in nails and bridges?

Student answers may vary. Both sources are metallic iron and are

attracted to magnets. They differ in form, purity, and size.

GENERAL CONCLUSIONS

1. **Designing Experiments** Use a metal file to scrape off metal flakes from a piece of metal or a nail. Compare this iron to the type collected from cereals.

The metal flakes will look similar, however the metal scrapings

will be larger.

Curdling the Bio-Tech Way

OBJECTIVES

Recommended time:
40 minutes

- **Test** and **observe** the curdling effect of rennet and chymosin.
- **Collect** and **interpret** data during milk curdling.
- **Compare** and **evaluate** the effectiveness of rennet and chymosin in cheese-making.

INTRODUCTION

Cheese-making is a complex biochemical process. First the milk is pasteurized to destroy any harmful bacteria. The milk is cooled, and then a special blend of bacteria or an acidic solution, such as vinegar or lemon juice, is added to the milk. The milk is left for one-and-a-half hours, during which time the bacteria convert lactose (milk sugar) into lactic acid, causing the milk to sour and the pH to drop to approximately 4.5. Rennet is then added to the milk, and within a short time, curds and whey are produced. Rennet, a group of enzymes, breaks down the milk protein casein to form paracasein. Paracasein combines with calcium found in milk to form paracaseinate, which separates out. Milk fat and water combine with paracaseinate to form curds. The remaining liquid is the whey. After curdling, the whey is removed and the curds are processed into cheese.

Until recently, rennet was extracted from the lining of the fourth stomach of milk-fed calves, where their mother's milk is digested. Scientists have identified chymosin as the most important enzyme in rennet, and they have developed a method to produce chymosin without the use of calves. To make chymosin genetically, scientists extract from calf stomach cells the DNA that contains the instructions for producing chymosin molecules. The DNA is inserted into bacteria or fungi. As the microorganisms grow, they produce large amounts of chymosin, which is purified before use. Currently, over 70% of the cheese made in the United States is coagulated with chymosin.

Genetically engineered chymosin was approved by the Food and Drug Administration (FDA) in March 1990. The FDA has determined that the genetically engineered chymosin is safe because it has the same structure and function as the animal-derived chymosin. The manufacturing process removes most impurities, and the host microorganisms, which are not pathogenic to humans, are destroyed or removed during processing. Still, many people question the safety of foods derived by biotechnology techniques.

In the following experiment, you will compare the curdling activities of animal-derived rennet with genetically engineered chymosin.

SAFETY

Solution/Material Preparation

1. To prepare rennet solution from a tablet, dissolve one tablet in 250 mL of water or according to the directions provided by the manufacturer.

2. Set out the milk containers before class so the milk will be at room temperature for the experiment.

3. Select small, oblong, disposable aluminum pans used for baking. These pans can be found in many grocery stores. Pot-pie tins could also be used, but the larger pie plates can be easily knocked off the hot plate.

  Always wear safety goggles and a lab apron to protect your eyes and clothing. If you get a chemical in your eyes, immediately flush the chemical out at the eyewash station while calling to your teacher. Know the location of the emergency lab shower and the eyewash station and the procedure for using them.

 Do not touch any chemicals. If you get a chemical on your skin or clothing, wash the chemical off at the sink while calling to your teacher. Make sure you carefully read the labels and follow the precautions on all containers of chemicals that you use. If there are no precautions stated on the label, ask your teacher what precautions you should follow. D not taste any chemicals or items used in the laboratory. Never return leftovers to their original containers; take only small amounts to avoid wasting supplies.

 Call your teacher in the event of a spill. Spills should be cleaned up promptly, according to your teacher's directions.

 Never put broken glass in a regular waste container. Broken glass should be disposed of properly.

Never stir with a thermometer because the glass around the bulb is fragile and might break.

 Do not heat glassware that is broken, chipped, or cracked. Use tongs or a hot mitt to handle heated glassware and other equipment because hot glassware does not always look hot.

MATERIALS

Rennet is available in tablet or liquid form. The solid is much more potent than the liquid. Junket rennet, a collection of protease enzymes, can be used but is not the same enzymes used to make cheese.

Use vinegar or lemon juice from the grocery store to acidify the milk.

- chymosin
- whole milk
- rennet tablets
- vinegar or lemon juice
- 50 mL beakers, 2
- 250 mL beaker
- 50 mL graduated cylinder
- 10 mL graduated cylinder
- aluminum foil pan, oblong
- balance
- cheesecloth
- funnel
- hot plate
- medicine droppers, 2
- stopwatch or watch with a second hand
- thermometer, nonmercury
- tongue depressors or popsicle sticks, 2
- universal pH paper
- watch glass or weighing boat
- water bath
- wax pencil

PROCEDURE

Techniques to Demonstrate

• You may wish to demonstrate the filter setup, especially how to line the funnel with cheesecloth and how to squeeze the last of the liquid from the curds.

Procedural Tips

• The curd is the most firm when one drop of chymosin is used per 50 mL of milk.
• Whole milk works best, but powdered milk or milk with less fat content will also work.

Pre-lab Discussion

• Students should be familiar with the structure of proteins and with the pH scale.
• Discuss with students the power of the enzymes used in the experiment. One drop of chymosin can coagulate up to 200 mL of milk. One part of commercial rennet can coagulate 5000 parts of milk.

Part 1: Making cheese, or forming curds and whey

1. Use a wax pencil to label two 50 mL beakers 1 and 2. Fill a small aluminum pan about 1 in. deep with tap water. Set the pan on a hot plate, and bring the temperature of the water to 37°C. Adjust the heat so that the water remains between 37°C and 40°C throughout the experiment.

2. To each beaker, add 30 mL of whole pasteurized milk at room temperature. Then add 1 mL of lemon juice or vinegar to each beaker. Gently swirl each beaker to mix the two liquids together. Using a pH paper strip, determine the pH of each milk solution. Record your results in **Data Table 1.**

3. Place both beakers in the warm water bath for 1 minute.

4. To the milk mixture in beaker 1, add 1 drop of chymosin. To the milk mixture in beaker 2, add 1 drop of rennet solution. Swirl each beaker gently for about 10 seconds to mix.

5. After swirling, return the beakers to the water bath. Let them sit undisturbed for 2 minutes. Record in **Data Table 1** the time needed for curds to start forming in each beaker.

6. After the 2-minute incubation period, gently tilt each beaker to observe the firming curd. Record your observations in **Data Table 1.**

7. Remove the beakers from the water bath. Using separate tongue depressors or popsicle sticks, cut the curd formed in each beaker into a tic-tac-toe pattern. Return the beakers to the water bath for 1 minute. During this time, the curd will start to contract and release the whey.

8. Rest a funnel on the rim of a 250 mL beaker as shown in **Figure A.** Line the funnel with several layers of cheesecloth. Pour the contents of beaker 1 through the funnel lined with cheesecloth to separate the curds from the whey. Allow the whey to drain into the collection beaker for 2–3 minutes. Then twist the top of the cheesecloth closed, and gently squeeze the whey from the cheese. Repeat the filtering process to separate the curds and whey in beaker 2.

Curds

Cheesecloth for filtering material

Funnel

250 mL beaker

Whey

FIGURE A

Disposal

Solutions can be poured down the drain. Cheese solids should be collected together, placed in a transparent plastic bag, labeled "Not for Consumption," and disposed of with the trash.

Part 2: Analyzing the curd and whey

9. Using a graduated cylinder, measure the volume of the whey collected from each beaker. Record the amounts in **Data Table 2**. Determine the pH of the whey, using a pH paper strip. Record the results in **Data Table 2.**

10. Open the cheesecloths. Use separate tongue depressors or popsicle sticks to transfer the curd to a pre-massed watch glass or weighing boat. Determine the mass of the curds formed in each beaker to the nearest 0.1g. Record these values in **Data Table 2**. Compare the average size, firmness, and texture of both cheese curds. Record your results in **Data Table 2.**

Cleanup and Disposal

11. Clean all apparatus and your lab station. Return equipment to its proper place. Dispose of chemicals and solutions in the containers designated by your teacher. Do not pour any chemicals down the drain or put them in the trash unless your teacher directs you to do so. Wash your hands thoroughly after all work is finished and before you leave the lab.

Data Table 1—*Curd Formation*

Variable	Beaker 1 (chymosin)	Beaker 2 (rennet)
pH of milk solution	4.6 or less	4.6 or less
Time of curd formation (s)	80 s	95 s
General observations	Accept all reasonable answers.	Accept all reasonable answers.

Data Table 2—*Curd and Whey Analysis*

Variable	Beaker 1 (chymosin)	Beaker 2 (rennet)
Volume of whey (mL)	170 mL	175 mL
pH of whey	4.5	4.5
Mass of curds (g)	20 g	18 g
Average size of curds (cm)	Answers will vary depending on the enzyme brand used and concentration.	
Firmness of curd	Accept all reasonable answers.	Accept all reasonable answers.
Texture of curd	Accept all reasonable answers.	Accept all reasonable answers.

QUESTIONS

1. Evaluating Data Describe any differences between chymosin and rennet activity.

Answers will vary. Typically, chymosin acts faster because it is the

active ingredient in rennet and is more pure.

2. Analyzing Data Were there any measurable differences in rates of curd formation between rennet and chymosin? Justify your answer.

Answers may vary, depending on the starting concentration of

the enzymes used. Check that data support the response.

3. Evaluating Data Describe any differences in size, texture, or firmness of the curds formed.

Answers will vary, however, there should not be any noticeable

differences with respect to curd size, texture, or firmness.

4. Evaluating Methods Why is it necessary to warm the milk to 37°C? *Hint:* Think of body temperature and enzyme activity in our bodies.

The optimum activity temperature for these enzymes is 37°C.

5. Evaluating Methods Why is it necessary to acidify the milk prior to the addition of the milk-digesting enzymes?

Acidification aids in curd formation and is essential in the produc-

tion of good cheese. If the pH in the milk is high, the curd

formed will be pasty; if the pH is low, the curd will be crumbly.

GENERAL CONCLUSIONS

1. **Predicting Outcomes** How would diluting the enzyme affect its ability to form curds? Will curd formation occur if you were to use a very dilute enzyme solution?

The time needed to form curds would increase because there is

less enzyme available. Curd formation would probably occur if

one waited long enough.

2. **Designing experiments** Design an experiment that tests the effect of milk fat on curdling.

Answers will vary but should include using milk samples of vary-

ing fat content (1%, 2%, skim, powdered, condensed, or whole

milk), a method and source (rennet or chymosin) for curdling,

and a defined means for collecting and comparing data.

3. **Designing experiments** Design an experiment that investigates the effect of temperature on curdling. How could this experiment be used to determine the temperature range that best promotes curd formation?

Accept all reasonable answers.

Name_____

Date_____ Class_____

EXPERIMENT **D22**

Electric Charge

OBJECTIVES

Recommended time:
30 min

INTRODUCTION

Required Precautions

• Safety goggles and a lab apron must be worn at all times.

SAFETY

• In case water splashes, use paper towels to mop up the water. Then at the sink, wring them out until they are only damp, and put them in the trash.

MATERIALS

PROCEDURE

Procedural Tips

• Students should be aware that high humidity makes charging the comb more difficult. They should also be aware that hair creams or hair spray may inhibit charge transfer.
• You may want to suggest that students try a variety of combs.

• **Observe and describe** the effects of static charge on a stream of water.

• **Infer** explanations for observed behaviors of the water stream.

When hair and nylon are rubbed together, electrons are transferred between them. Both the hair and the comb become charged. When the charged comb comes close to the water stream, it induces a complementary charge on the water stream, causing it to move toward the comb. In this activity, you will attract a stream of water to a charged nylon comb.

  Always wear safety goggles and a lab apron to protect your eyes and clothing. If you get a chemical in your eyes, immediately flush the chemical out at the eyewash station while calling to your teacher. Know the location of the emergency lab shower and the eyewash station and the procedure for using them.

• nylon hair comb
• faucet with cold running water
• 15 cm plastic ruler

1. Turn on a cold-water faucet. Adjust the stream to a diameter of about 1.5 mm. Make sure the stream is a steady flow.

2. Pull a comb through your hair several times. Be sure that the comb does not get wet.

3. Holding the comb about 10 cm below the faucet, slowly bring the teeth of the comb within 3 cm of the water stream. Observe what happens.

Observations:

The stream of water should bend toward the nylon comb.

_____

_____

_____

_____

4. Move the comb closer to the water. Observe what happens.

Observations:

The bending of the stream should increase.

5. Repeat steps 2 through 4. Note any changes in your initial observations.

Observations:

Answers will vary, but students should observe a more strongly

bent stream because of the increased static charge.

6. Adjust the faucet to produce a larger stream of water. Repeat steps 2 through 4.

Observations:

The water stream does not bend.

Cleanup and Disposal

7. Clean the comb and your lab station. Return equipment to its proper place. Wash your hands thoroughly after all work is finished and before you leave the lab.

QUESTIONS

1. Analyzing Results How does the distance between the comb and the water stream affect the results?

The smaller the distance between the comb and the stream, the

more the stream bends.

2. Applying Ideas How does a second charging of the comb affect the bending of the water stream? Give a reason for the change in behavior of the water stream.

Presuming the comb has residual charge when the hair is stroked

the second time, the water stream will bend more the second

time than the first time. A large static charge induces a stronger

complementary charge on the water stream.

3. **Analyzing Results** When the size of the water stream is increased, does the water bend more or less than originally? Explain your observation.

Less; the charge induced by the comb must be spread over more

water molecules, so the effect on any one molecule is consider-

ably less than in the case of the thin stream.

_____

_____

_____

GENERAL CONCLUSIONS

1. **Applying Ideas and Predicting Outcomes** Water is an electrically neutral polar molecule. Consider the structures of isopropyl alcohol and carbon tetrachloride. If the stream of liquid were isopropyl alcohol or carbon tetrachloride instead of water, would the results of the experiment be the same as the results obtained with water? Explain your reasoning.

No. Isopropyl alcohol molecules are electrically neutral and only

slightly polar, so the static charge on the comb may bend the al-

cohol stream ever so slightly but not as much as water. Carbon

tetrachloride molecules are electrically neutral and nonpolar. The

static charge on the comb will not be strong enough to induce a

stream of CCl_4 to bend.

_____

_____

_____

2. **Predicting Outcomes** When clothes are dried in an electric dryer, a dryer sheet is sometimes added to reduce static cling and to give the clothes a fresh smell. Suppose the comb in this experiment is rubbed with a dryer sheet before it is drawn through the hair. Predict the effect on the water stream when the comb is brought near it. Explain your reasoning.

The water stream should not bend because sizing from the dryer

sheet has coated the comb and prevented a sufficient transfer of

electrons between the hair and comb.

_____

_____

3. **Inferring Results and Designing Experiments** How would the sign of the charge (positive or negative) affect the response of the water stream? Design an experiment to test your hypothesis.

The water stream should bend for either type of charge, but the

direction should change. Answers for the experiment will vary.

Student responses should include a source and method for pro-

ducing a positively charged substance and a negatively charged

substance and should also indicate how changes in the water

stream will be detected.

4. **Applying Concepts** During dry winter seasons, static cling is prevalent, affecting clothes and hair. Applying hair spray to a comb before pulling it through the hair can reduce or prevent static cling. Explain.

Answers will vary, but students should recognize that the coated

surface does not transfer electrons as readily as the noncoated

surface, so the concentration of charge is not sufficient to pro-

duce cling.